互联网＋珠宝系列教材
高等教育珠宝专业"十三五"规划教材

珠宝专业英语

（第二版）

Jewelry Professional English

主　编　肖启云
副主编　姚满林　石振荣　刘衔宇

中国地质大学出版社
CHINA UNIVERSITY OF GEOSCIENCES PRESS

前 言

（第一版）

我国即将成为全球最具竞争力的珠宝首饰制造和贸易中心之一，也将成为世界最大的珠宝消费市场，珠宝首饰业的全球化发展要求鉴定、评估、商贸等从业者具有较高的专业外语水平。为了使学生及现有珠宝首饰从业者较快地适应这一新形势，我们编写了这本教材。

珠宝专业英语是综合珠宝专业知识和英语运用能力的课程，故在体系设置时我们兼顾了英语教学和珠宝专业教学的规律。通过本课程的系统学习，可以掌握珠宝首饰销售过程中所必备的英语会话和交流技巧；掌握常见的珠宝首饰专业词汇；提高阅读及理解珠宝专业英文资料的能力；掌握珠宝首饰专业资料翻译的方法和技巧，从而具备在珠宝首饰相关工作中解决与专业英语有关问题的能力。

全书所有材料取自于原版英文书刊或专业网站，内容丰富、题材广泛、语言流畅、文字活泼、图文并茂、条理清楚、通俗易懂、实用性强，其特色如下：

1. 内容全面：涵盖了宝石学科各方面的内容，共有16课，包括宝石的概念与分类、光学性质、力学性质、鉴定仪器、合成与优化处理、包裹体、鉴定步骤、加工、重要宝石和宝石商贸等内容；

2. 每课各部分内容循序渐进、由浅入深，如从珠宝销售对话开始，依次为专业词汇训练、关键短句、短文、课文、词汇表和练习题；

3. 以实用为目的，将口语交流与掌握专业词汇融为一体，强调语言技能和专业知识并重，提高珠宝专业英语的实际运用能力；

4. 附有注音标的专业词汇总表、宝石名称的中英文对应表，便于读者查询，可兼作珠宝专业工具书之用；

5. 附配套的新西兰珠宝专业人士录音的光盘，使读者可以边听边学，既便于记忆又得到语音训练；

6. 附配套的参考译文，供师生上课或个人学习参考。

教材第一课至第八课的中文由中国地质大学(北京)博士生凌潇潇女士翻译，编者在此表示诚挚的谢意。

侨居在新西兰的李海宏先生和新西兰珠宝专业 Anne 女士一起对教材进行了通篇的校对，对一些词句的表达再三斟酌修改，使得各语句更贴近现今英语使用习惯，在此特表谢意！

配套的光盘中，各课对话部分由新西兰籍珠宝专业 Anne 女士和 Rachel 女士朗读，其他各部分由 Anne 女士朗读，光盘内容由清华大学计算机专业工程师姚满林先生编辑处理，在此一并致谢！

此外，该教材的编写得到了北京城市学院的大力支持，在此表示感谢！

编者在编写过程中参阅或选用了许多国内外相关著作、教材、网站资料，在此对各原作者表示衷心的感谢！

由于编者水平有限，经验不足，敬请专家、学者不吝指正。

如需了解本书相关信息或提出建议，请与编者联系（xiaoqiyun@sohu.com）。

编者

2011 年 4 月

目 录

Unit 1　Gems and Gemstones Instruction ……………………………………… (1)

　　Part 1　Dialogue ………………………………………………………………… (1)

　　Part 2　Fill the Chinese Meanings with the Teacher's Tutorship ………… (2)

　　Part 3　Link the Words and the Relevant Pictures ………………………… (2)

　　Part 4　Useful Phrases ………………………………………………………… (2)

　　Part 5　Some Sentences ……………………………………………………… (3)

　　Part 6　Short Paragraph ……………………………………………………… (3)

　　Part 7　Text ……………………………………………………………………… (3)

　　Part 8　Words and Expressions ……………………………………………… (7)

　　Part 9　Check Your Understanding ………………………………………… (10)

Unit 2　Optical Properties of Cut Gemstones ……………………………… (11)

　　Part 1　Dialogue ……………………………………………………………… (11)

　　Part 2　Fill the Chinese Meanings with the Teacher's Tutorship ……… (12)

　　Part 3　Link the Words and the Relevant Pictures ……………………… (12)

　　Part 4　Useful Phrases ……………………………………………………… (12)

　　Part 5　Some Sentences ……………………………………………………… (13)

　　Part 6　Short Paragraph ……………………………………………………… (13)

　　Part 7　Text ……………………………………………………………………… (15)

　　Part 8　Words and Expressions ……………………………………………… (19)

　　Part 9　Check Your Understanding ………………………………………… (21)

Unit 3　Physical Properties of Gems ………………………………………… (23)

　　Part 1　Dialogue ……………………………………………………………… (23)

　　Part 2　Fill the Chinese Meanings with the Teacher's Tutorship ……… (24)

	Part 3	Link the Words and the Relevant Pictures	(24)
	Part 4	Useful Phrases	(24)
	Part 5	Some Sentences	(25)
	Part 6	Short Paragraph	(25)
	Part 7	Text	(26)
	Part 8	Words and Expressions	(29)
	Part 9	Check Your Understanding	(31)

Unit 4 Gemological Instruments (32)

	Part 1	Dialogue	(32)
	Part 2	Fill the Chinese Meanings with the Teacher's Tutorship	(33)
	Part 3	Link the Words and the Relevant Pictures	(33)
	Part 4	Useful Phrases	(33)
	Part 5	Some Sentences	(34)
	Part 6	Short Paragraph	(34)
	Part 7	Text	(35)
	Part 8	Words and Expressions	(39)
	Part 9	Check Your Understanding	(42)

Unit 5 Synthetic Gems Introduction (43)

	Part 1	Dialogue	(43)
	Part 2	Fill the Chinese Meanings with the Teacher's Tutorship	(44)
	Part 3	Link the Relevant Words between Column A and Column B	(44)
	Part 4	Useful Phrases	(44)
	Part 5	Some Sentences	(45)
	Part 6	Short Paragraph	(45)
	Part 7	Text	(46)
	Part 8	Words and Expressions	(49)
	Part 9	Check Your Understanding	(50)

Unit 6 Gemstone Treatments and Enhancements (52)

	Part 1	Dialogue	(52)
	Part 2	Fill the Chinese Meanings with the Teacher's Tutorship	(53)
	Part 3	Link the Relevant Words between Column A and Column B	(53)

	Part 4	Useful Phrases	(53)
	Part 5	Some Sentences	(54)
	Part 6	Short Paragraph	(54)
	Part 7	Text	(55)
	Part 8	Words and Expressions	(59)
	Part 9	Check Your Understanding	(62)
	Part 10	Self-Study Material	(63)

Unit 7 Gemstone Inclusion (64)

	Part 1	Dialogue	(64)
	Part 2	Fill the Chinese Meanings with the Teacher's Tutorship	(65)
	Part 3	Link the Relevant Words between Column A and Column B	(65)
	Part 4	Useful Phrases	(65)
	Part 5	Some Sentences	(65)
	Part 6	Short Paragraph	(66)
	Part 7	Text	(67)
	Part 8	Words and Expressions	(70)
	Part 9	Check Your Understanding	(72)
	Part 10	Self-Study Material	(73)

Unit 8 Gemstones Identification Procedure (75)

	Part 1	Dialogue	(75)
	Part 2	Fill the Chinese Meanings with the Teacher's Tutorship	(76)
	Part 3	Link the Words and the Relevant Pictures	(76)
	Part 4	Useful Phrases	(76)
	Part 5	Some Sentences	(77)
	Part 6	Short Paragraph	(77)
	Part 7	Text	(78)
	Part 8	Words and Expressions	(80)
	Part 9	Check Your Understanding	(82)
	Part 10	Self-Study Material	(83)

Unit 9 Gemstone Cut (85)

	Part 1	Dialogue	(85)

Part 2	Fill the Chinese Meanings with the Teacher's Tutorship	(86)
Part 3	Link the Words and the Relevant Pictures	(86)
Part 4	Useful Phrases	(86)
Part 5	Some Sentences	(87)
Part 6	Short Paragraph	(87)
Part 7	Text	(89)
Part 8	Words and Expressions	(90)
Part 9	Check Your Understanding	(91)
Part 10	Self – Study Material	(92)

Unit 10 Diamond (94)

Part 1	Dialogue	(94)
Part 2	Fill the Blanks Basing on Your Gemological Knowledge	(95)
Part 3	Link the Words and the Relevant Pictures	(95)
Part 4	Useful Phrases	(95)
Part 5	Some Sentences	(96)
Part 6	Short Paragraph	(96)
Part 7	Text	(98)
Part 8	Words and Expressions	(100)
Part 9	Check Your Understanding	(102)
Part 10	Self – Study Material	(103)

Unit 11 Ruby and Sapphire (105)

Part 1	Dialogue	(105)
Part 2	Fill the Blanks Basing on Your Gemological Knowledge	(106)
Part 3	Link the Words and the Relevant Pictures	(106)
Part 4	Useful Phrases	(106)
Part 5	Some Sentences	(107)
Part 6	Short Paragraph	(107)
Part 7	Text	(108)
Part 8	Words and Expressions	(110)
Part 9	Check Your Understanding	(112)
Part 10	Self – Study Material	(113)

Unit 12	Emerald	(114)
Part 1	Dialogue	(114)
Part 2	Fill the Blanks Basing on Your Gemological Knowledge	(115)
Part 3	Link the Words and the Relevant Pictures	(115)
Part 4	Useful Phrases	(115)
Part 5	Some Sentences	(116)
Part 6	Short Paragraph	(116)
Part 7	Text	(117)
Part 8	Words and Expressions	(119)
Part 9	Check Your Understanding	(120)
Part 10	Self-Study Material	(120)

Unit 13	Quartz Gemstone	(122)
Part 1	Dialogue	(122)
Part 2	Fill the Blanks Basing on Your Gemological Knowledge	(123)
Part 3	Link the Words and the Relevant Pictures	(123)
Part 4	Useful Phrases	(123)
Part 5	Some Sentences	(124)
Part 6	Short Paragraph	(124)
Part 7	Text	(125)
Part 8	Words and Expressions	(127)
Part 9	Check Your Understanding	(129)
Part 10	Self-Study Material	(130)

Unit 14	Pearl	(131)
Part 1	Dialogue	(131)
Part 2	Fill the Blanks Basing on Your Gemological Knowledge	(132)
Part 3	Link the Words and the Relevant Pictures	(132)
Part 4	Useful Phrases	(132)
Part 5	Some Sentences	(133)
Part 6	Short Paragraph	(133)
Part 7	Text	(135)
Part 8	Words and Expressions	(136)

Part 9	Check Your Understanding	(138)
Part 10	Self – Study Material	(138)

Unit 15　Jadeite Jade　(140)

Part 1	Dialogue	(140)
Part 2	Fill the Blanks Basing on Your Gemological Knowledge	(141)
Part 3	Link the Words and the Relevant Pictures	(141)
Part 4	Useful Phrases	(141)
Part 5	Some Sentences	(142)
Part 6	Short Paragraph	(143)
Part 7	Text	(143)
Part 8	Words and Expressions	(145)
Part 9	Check Your Understanding	(147)
Part 10	Self – Study Material 1	(147)
Part 11	Self – Study Material 2	(148)

Unit 16　Jewelry Commerce　(150)

Part 1	Dialogue	(150)
Part 2	Fill the Blanks Basing on Your Gemological Knowledge	(151)
Part 3	Link the Paragraphs and the Relevant Pictures	(151)
Part 4	Useful Phrases	(152)
Part 5	Some Sentences	(153)
Part 6	Short Paragraph	(153)
Part 7	Text	(155)
Part 8	Words and Expressions	(157)
Part 9	Check Your Understanding	(160)
Part 10	Self – Study Material	(160)

附录1　珠宝玉石英文单词表　(163)

附录2　化学元素(部分)中英文对照表　(169)

参考译文　(171)

主要参考文献　(208)

Unit 1　Gems and Gemstones Instruction

Part 1　Dialogue

本章音频

Grace：What can I do for you?
Kitty：I'd like some jewelry.
Grace：All the jewelry is on sale today.
Kitty：I'd like emerald ring and pearl necklaces.
Grace：Sure. Here is a nice pearl necklace.
Kitty：May I have a look?
Grace：Yes, why not have a look at the nephrite bracelet and amethyst pendant by the way?
Kitty：The necklace is very elegant. I'll take it. What is that?
Grace：The gemstone? Oh, it's tourmaline and the chain is 14K gold.

Kitty：What's the price?
Grace：Its regular price is ＄880, and now you can have it with a twenty percent discount.
Kitty：How about six hundred dollars?

Grace：I'm sorry we only sell at fixed prices.
Kitty：Oh, I'll take the pearl necklace and that chain with fancy tourmaline.

Part 2　Fill the Chinese Meanings with the Teacher's Tutorship

ruby _____   diamond _____   tourmaline _____   emerald _____
pearl _____   sapphire _____   nephrite _____   coral _____
jadeite _____ amethyst _____   amber _____   opal _____

Part 3　Link the Words and the Relevant Pictures

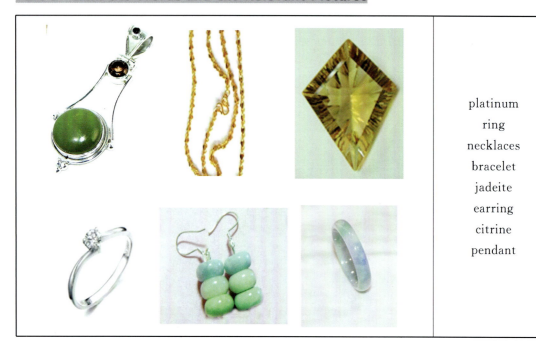

platinum
ring
necklaces
bracelet
jadeite
earring
citrine
pendant

Part 4　Useful Phrases

golden content analysis instrument　黄金成色仪
diamond grading　钻石分级
stone sculpture　石雕,石刻艺术
historical tale on pearls　珍珠史话
18 karat gold jewelry　18K 金首饰
imitation diamond　仿钻
appreciation of opal　欧泊的质量评估
the rules for facet design　刻面宝石的设计原则
white jade　白玉
the origin of rings, necklaces and bracelets　戒指、项链和手镯的起源
plum blossom jade　梅花玉
agate with water　水胆玛瑙

platinum ornament　铂金饰品
synthetic gemstone　合成宝石
jewelry technology　珠宝首饰工艺

Part 5　Some Sentences

(1) Gemology is the study of gemstones, which includes gem properties, locations and origins. Gemology is often studied by people in the jewelry business, including business owners, buyers, designers and appraisers. Others studying gemology include antique dealers and auction house catalogers. People in these jobs need to be able to identify gems and gemstones and describe their properties.

(2) Guilin ruby crystal is synthesized by hydrothermal method. The crystal growing equipment and technological conditions are also studied. The determination on gemological characters has been done.

(3) This article introduces the polishing and drilling process of faceted quartz crystal. The characters of equipments and polishing materials are also discussed.

Part 6　Short Paragraph

Important Qualities of a Gemstone

A gemstone can be valued for its natural growth having the traditionally quoted "three cardinal virtues" of beauty, durability and rarity. Acceptability and portability are other factors decided by the society for valuation of gemstones.

Beauty is related to the results of the visible light by interaction with the gemstone. These effects are different for different types of gem materials, specifically in relation to the degree of transparency.

Durability is related to the structure, which depends on hardness against scratching, toughness against breaking and stability against external forces such as heat, pressure, shock, and chemical action etc.

Rarity of a gemstone depends on its availability, and the valuation is decided by the supply and demand theory in economics.

Part 7　Text

Gems and Gemstones Instruction

What Is a Gemstone? What Is a Gem?

Have you ever seen a diamond in a ring? An opal in a necklace? A pearl earring? Diamonds, opals, and pearls are types of gemstones. A gemstone is a mineral, rock, or organic material that is used for jewelry, ornamentation, or art. A gem, such as a diamond, is a type of gemstone that must be cut and polished for its beauty to be visible. Opals and pearls may be styled into jewelry or art without any cuts or polishing. They are gemstones, but not gems.

> *This pendant has cut and polished diamonds and an opal. Do you think the diamonds are gems or gemstones? What about the opal? Answer: Diamonds are gems, opal isn't.*

A gem is a natural, mineral or organic substance, that has substantial beauty, rarity and durability.

Natural means that the material was not made, or assisted in its making, by any human effort. When such is the case, modifiers such as "laboratory grown", "synthetic", "cultured", or "man-made", must, by Federal Trade Commission (FTC) regulations, be used in the descriptions of any such pieces being advertised or marketed. Man-made "gems" have all the chemical, optical and physical characteristics of the natural materials they imitate, but they do not have their rarity or value. You can be certain whenever you see any of the above modifiers that the material in question is not of natural origin.

A mineral can be defined as a crystalline solid with a specific chemical formula, and a regular three dimensional arrangement of atoms.

Where Are Gemstones Found?

Gemstones occur in locations all over the world. Diamonds are found deep within the earth in a rock called kimberlite. Tourmaline and beryl are found in stream beds after they erode from surrounding rocks. And garnet is often found in a rock called gneiss, which has been heated to high temperatures.

Tourmaline comes in many colors and sometimes one crystal can have multiple colors. The tourmaline crystals attached to the quartz are tri-colored which means they are made up of three colors.

How Are Gemstones Formed?

Gemstones are formed in several specific and different ways. Their colors are almost always a result of their chemical compositions while they form.

Turquoise forms when water moves through a rock containing copper, aluminum and phosphorus. Turquoise often occurs in arid or desert environments, such as the southwestern part of the United States. You may be familiar with beautiful native American jewelry from the southwest, which is often made with

turquoise.

Lapis lazuli is a rock and not a mineral. Lapis lazuli forms when magma under the surface of the earth forces its way into an existing rock. The magma is so hot that the existing rock melts and then solidifies. This process creates a new deep blue rock, lapis lazuli, which contains the minerals lazurite, pyrite and calcite.

People commonly think of garnets as being red, but they are found in many colors ranging from yellow to black. Color-changing garnets look different when viewed in daylight and incandescent light.

Garnets often form in hot metamorphic rocks under great pressure. Garnets occur in every color. Their color is determined by the chemical compositions of the melted mineral mix as it solidifies. Red garnets, or pyrope, get their color from magnesium silicate, the melted chemical mixture in which they form.

Jade is a highly valued material used in burial ceremonies, royal crowns, jewelry and for the hieroglyphics of many cultures such as the Olmecs of Mesoamerica and Chinese.

Jade is a gem that can be cut and polished from two minerals: jadeite and nephrite. Both minerals exist in metamorphic rocks deep within the earth.

Synthetic Gemstones and Gems

Gemstones and gems can be made in laboratories. Scientists try to create the same conditions in the laboratory as in the earth because similar conditions give the gemstones and gems similar properties. Turquoise, sapphires and rubies can be created in a synthetic environment. In one method for making rubies, a rod with a "seed crystal" is lowered into melted minerals and then brought back up. Repeating this process over and over grows a large crystal on the end of a rod from the melted minerals. The ruby can then be detached, cut and polished.

Gemstone Classification

Scientists and gemologists have developed a number of ways to classify gemstones: precious or semiprecious, natural or synthetic, and organic or inorganic.

The precious gem—Hope Diamond has been around for

centuries and has had many owners during its lifetime. The different owners have recut, polished and reset it for multiple times, helping shape what it is today.

Precious or Semiprecious?

This classification is based on beauty, rarity and hardness. It applies mainly to gems, rather than gemstones. The most beautiful, rarest, and hardest gems are considered as precious gems. Precious gems include diamonds, rubies, sapphires, emeralds, aquamarines, topazes and opals. A semiprecious gem is a gem that is less beautiful, less rare and less hard. It is also less than 8 on the Mohs' Hardness Scale, which means it is easier to scratch. Turquoise, jade, lapis lazuli and amber are all semiprecious.

A sapphire miner holds several, tiny gemstones in her hand that she collected from the mine in Madagascar.

Would you consider topaz more beautiful than turquoise? It's hard to say. Classifying gems as precious or semiprecious is fading in popularity because it is easy to disagree about beauty.

Natural or Synthetic?

A natural gemstone is one that is formed in the earth. A synthetic gemstone is made in the laboratory. Some gemstones are both. Emeralds, garnets, rubies, sapphires and diamonds can be both mined from the ground or made in a laboratory.

Have you ever seen a piece of amber with an insect or plant in it? This photo shows the early stages of how amber forms and how easily an ant can get stuck in the sticky tree sap.

Organic or Inorganic?

An organic gemstone is one that is created by a living thing. Pearls are created by oysters and mussels. Amber is sap created by a tree. Coral is created by tiny communities of animals in the ocean. Pearls, amber and coral are organic gemstones. Inorganic gems include diamonds, sapphires and rubies. These are created by minerals without the help of organisms.

Gemstones and People

1) Pearls in Culture

Hundreds of years ago, along the east coast of the United States, Native Americans collected pearls to use in jewelry. Both men and women wore pearl jewelry, including ear pendants with pearls. Pocahontas' father, Powhatan, reportedly had a collection of pearls given to him as a tribute.

2) Turquoise in Culture

Turquoise was mined in Persia for thousands of years. It was transported to other places, including Egypt, where it was used by the Pharaohs. Much later, turquoise was discovered in the southwestern United States. Now the United States is the largest producer of turquoise.

3) Quartz in Culture

Quartz is the second most common mineral in Earth's crust. There are many varieties of quartz including rose quartz, smoky quartz and amethyst.

Quartz is made of silicon and oxygen. Quartz is abundant on Earth and very hard. Because of these properties, it has been used for jewelry for at least 4000 years. Before quartz was used for jewelry, it was used for spear points. Some people believe quartz can be used for healing.

4) Opals in Culture

Ancient Greeks believed that opals gave the owner the power to see into the future. Romans believed that these gemstones were a symbol of purity. Arabs believed that opals fell from heaven. However, by the nineteenth century, many people believed that opals were associated with bad luck and should not be worn. Some people today still believe this. But, many others wear opals because they believe they are beautiful.

Part 8 Words and Expressions

gem[dʒem]	n. 宝石
gemstone['dʒem,stəʊn]	n. 经雕琢的宝石
ruby['ruːbɪ]	n. 红宝石
tourmaline['tʊəməlɪn]	n. 碧玺,电气石
emerald['emərəld]	n. 祖母绿
sapphire['sæfaɪə]	n. 蓝宝石
pearl[pɜːl]	n. 珍珠
coral['kɒrəl]	n. 珊瑚,珊瑚虫
nephrite['nefraɪt]	n. 软玉,和田玉
jadeite['dʒeɪdaɪt]	n. 翡翠
opal['əʊpl]	n. 欧泊,蛋白石
amethyst['æməθɪst]	n. 紫晶
amber['æmbə]	n. 琥珀
platinum['plætɪnəm]	n. 铂(元素符号 Pt)

ring[rɪŋ]	n. 戒指
necklace[ˈnekləs]	n. 项链
bracelet[ˈbreɪslət]	n. 手镯
citrine[ˈsɪtrɪn]	n. 黄晶
earring[ˈɪərɪŋ]	n. 耳环，耳饰
pendant[ˈpendənt]	n. 垂饰，吊坠
sculpture[ˈskʌlptʃə]	n. 雕刻，雕刻品，雕塑，雕塑品　v. 雕刻，雕塑，刻蚀
grading[ˈgreɪdɪŋ]	n. 分级
karat[ˈkærət]	n. (=carat)克拉(宝石的重量单位)
imitation[ˌɪmɪˈteɪʃn]	n. 赝品，仿造物
appreciation[əˌpriːʃɪˈeɪʃn]	n. 增值，欣赏，鉴赏，正确评价，感谢，感激
facet[ˈfæsɪt]	n.(宝石等的)刻面，小平面，方面，琢面，(多面体的)面 vt. 在……上刻画
jade[dʒeɪd]	n. 翡翠
plum blossom	梅花
agate[ˈægət]	n. 玛瑙
ornament[ˈɔːnəmənt]	n. 装饰物
synthetic[sɪnˈθetɪk]	adj. 合成的，人造的，综合的
synthesize[ˈsɪnθəsaɪz]	v. 综合，合成
gemology[dʒeˈmɒlədʒɪ]	n. 宝石学
property[ˈprɒpətɪ]	n. 性质，特性
location[ləʊˈkeɪʃn]	n. 位置，场所，特定区域
appraiser[əˈpreɪzə]	n. 评价者
antique[ænˈtiːk]	n. 古物，古董　adj. 古时的，过时的
dealer[ˈdiːlə]	n. 经销商，商人
auction[ˈɔːkʃn]	n. 拍卖
identify[aɪˈdentɪfaɪ]	vt. 识别，鉴别
crystal[ˈkrɪstl]	n. 晶体，水晶
hydrothermal[ˌhaɪdrəʊˈθɜːməl]	adj. 热水的，热液的
polish[ˈpɒlɪʃ]	vt. 擦亮，发亮，磨光，推敲　vi. 发亮，变光滑
drill[drɪl]	v. 训练，钻孔，条播
quartz[kwɔːts]	n. 石英
durability[ˌdjʊərəˈbɪlətɪ]	n. 经久，耐久性
rarity[ˈreərətɪ]	n. 稀有
transparency[trænsˈpærənsɪ]	n. 透明，透明度
hardness[ˈhɑːdnəs]	n. 硬，硬度
scratch[skrætʃ]	vt. 擦，刮
toughness[ˈtʌfnəs]	n. 韧性，坚韧，刚性，健壮性
availability[əˌveɪləˈbɪlətɪ]	n. 可用性，有效性，实用性

Unit 1 Gems and Gemstones Instruction

mineral[ˈmɪnərəl]	n. 矿物,矿石
organic[ɔːˈgænɪk]	adj. 有机的
kimberlite[ˈkɪmbəlaɪt]	n. 角砾云橄岩,金伯利岩(含金刚石)
beryl[ˈberəl]	n. 绿柱石
stream bed	河床
garnet[ˈgɑːnɪt]	n. 石榴石,深红色
gneiss[naɪs]	n. 片麻岩
turquoise[ˈtɜːkwɔɪz]	n. 绿松石
copper[ˈkɒpə]	n. 铜(元素符号 Cu)
aluminum[əˈljuːmɪnəm]	n. 铝(元素符号 Al)
phosphorus[ˈfɒsfərəs]	n. 磷(元素符号 P)
arid[ˈærɪd]	adj. 干旱的,贫瘠的(土地等)
lapis lazuli[ˈlæpɪs][ˈlæzjʊlaɪ]	n. 天青石,青金石,天青石色
magma[ˈmægmə]	n. (有机物或矿物的)稀糊,岩浆,乳浆剂
lazurite[ˈlæzjʊˌraɪt]	n. 天青石,青金石
pyrite[ˈpaɪraɪt]	n. 黄铁矿
calcite[ˈkælsaɪt]	n. 方解石
incandescent[ˌɪnkænˈdesnt]	adj. 遇热发光的,白炽的
pyrope[ˈpaɪrəʊp]	n. 镁铝榴石
magnesium[mægˈniːziəm]	n. 镁(元素符号 Mg)
silicate[ˈsɪlɪkət]	n. 硅酸盐
hieroglyphic[ˌhaɪərəˈglɪfɪk]	n. 象形文字
Olmec[ˈɒlmɛk]	n. 奥尔梅克人(居住在墨西哥东南部的有高度文化的古印第安人),奥尔麦克文化的
Mesoamerica[ˌmezəʊəˈmerɪkə]	n. 中亚美利加洲,中美洲
gemologist[dʒeˈmɒlədʒɪst]	n. 宝石学家
classification[ˌklæsɪfɪˈkeɪʃn]	n. 分类,分级
aquamarine[ˌækwəməˈriːn]	n. 海蓝宝石
topaz[ˈtəʊpæz]	n. 托帕石
oyster[ˈɔɪstə]	n. 牡蛎
mussel[ˈmʌsl]	n. 贻贝,蚌类
community[kəˈmjuːnɪtɪ]	n. 公社,社会,(政治)共同体,一致,(生物)群落
organism[ˈɔːgənɪzəm]	n. 生物体
tribute[ˈtrɪbjuːt]	n. 贡品,礼物,颂词,殷勤,贡物
mine[maɪn]	n. 矿,矿山
Persia[ˈpɜːʃə]	n. 波斯(西南亚国家,现在的伊朗)
Pharaoh[ˈfeərəʊ]	n. 法老王(古埃及君主称号)
crust[krʌst]	n. 外壳,硬壳
healing[ˈhiːlɪŋ]	n. 康复,复原

Part 9 Check Your Understanding

Exercise 1　Answer the following questions

(1) What does gemology relate to?
(2) How are the colors of gemstones formed?
(3) What is the synthetic gemstones?
(4) Please introduce the classification of gemstones.
(5) Give some examples about the jewelry culture.

Exercise 2　Fill in the blanks

(1) Please illustrate several kinds of gems. _____, _____ and _____.
(2) The noble metals such as _____ and _____ are often to be forged into ornaments.
(3) _____ is the scholar who studied the gems and gemstones.
(4) _____, _____ and _____ is the gem materials' necessary factor.
(5) The beautiful ruby is possibly synthesized by _____ method.

Exercise 3　Translation

(1) A gemstone is a mineral, rock, or organic material that is used for jewelry, ornamentation, or art. A gem, such as a diamond, is a type of gemstone that must be cut and polished for its beauty to be visible. Opals and pearls may be styled into jewelry or art without any cuts or polishing.
(2) Tourmaline comes in many colors and sometimes one crystal can have multiple colors.
(3) The most beautiful, rarest, and hardest gems are considered as precious gems. Precious gems include diamonds, rubies, sapphires, emeralds, aquamarines, topazes and opals.
(4) 蓝宝石是一种名贵宝石,但很难说蓝宝石就比托帕石漂亮。
(5) 市场上所见的钻石、红宝石、祖母绿、紫晶等可能是天然的,也可能是人工合成的。
(6) 红色的石榴石被切割成刻面后再抛光可以变得非常漂亮。

Unit 2　Optical Properties of Cut Gemstones

Part 1　Dialogue

Grace: Do you know why the gemstones are beautiful?
Kitty: The beauty of gems depends to a large extent on their optical properties, I think.
Grace: Explain them in detail?
Kitty: The color, the luster, the transparency, the dispersion, pleochroism etc.
Grace: Oh, I see. I remember that the teacher spoke of the optical phenomena.
Kitty: Do you refer to the chatoyancy, asterism and color change?
Grace: Yes, chrysoberyl is attractive because of its cat's eye, isn't it?
Kitty: Of course, alexandrite is costly by its color change in the same way.
Grace: Do you hear of opal?
Kitty: Yes, I have ever seen iridescent opal in Australia.
Grace: Come on, have a look at this opal ring I bought yesterday.

本章音频

Kitty: Oh, what a wonderful jewelry!
Grace: Take it, I bought it for you as birthday present.
Kitty: Thank you very much!

Part 2　Fill the Chinese Meanings with the Teacher's Tutorship

optical _____　　luster _____　　transparency _____　　dispersion _____
chatoyancy _____　asterism _____　chrysoberyl _____　　alexandrite _____
iridescent _____　　play of color _____　aventurization _____

Part 3　Link the Words and the Relevant Pictures

play of color
fire
chatoyancy
color change
asterism
moon
phenomena

Part 4　Useful Phrases

a guide for gem mounting　宝石镶嵌指南
identification of two different types of chicken-blood stone　两种鸡血石的鉴别
synthetic spinel's gemological characters　合成尖晶石的宝石学特征
irradiated gem ornament　辐照改色的宝石饰品
wedding jewelry　结婚首饰
assessment of ornament rock fluorite　萤石工艺品的评价
red coral: the treasure from deep seas　海中瑰宝红珊瑚
natural emerald　天然祖母绿
jewelry circle　珠宝界
jadeite expert　翡翠专家
synthetic moissanite　合成碳化硅
B-jadeite aging test conducted and its service life probed　B货翡翠老化检测及寿命探讨
gem and jewelry assessment trade　珠宝首饰评估业
appreciation and analysis of the jade-ware treasures　玉器珍品赏析
moonstone, amazonite, sunstone and labradorite　月光石、天河石、日光石和拉长石

Part 5 Some Sentences

(1) The beauty of gems depends to a large extent on their optical properties. The most important optical properties are the degree of refraction and color. Other properties include fire, the display of prismatic colors; dichroism, the ability of some gemstones to present two different colors when viewed in different directions; and transparency.

(2) Diamond is highly prized because of its fair and brilliancy, ruby and emerald because of the intensity and beauty of their colors, and star sapphire and star ruby because of the star effect, known as asterism, as well as for their color.

(3) In certain gemstones, notably opals, brilliant areas of color can be seen within the stone; These areas change in hue and size as the stone is moved. This phenomenon, known as play of color, differs from fire and is caused by interference and reflection of the light by tiny irregularities and cracks inside the stone.

(4) The appearance of a gem as seen by reflected light is another optical property of gemstones and is called luster. The luster of gems is characterized by the terms metallic, adamantine (like the luster of the diamond), vitreous (like the luster of glass), resinous, greasy, silky, pearly, or dull. Luster is particularly important in the identification of gemstones in their uncut state.

Part 6 Short Paragraph

Optical Phenomena in Gemstones

Optical phenomena encompass the light-dependent properties of a gem, which are not due to its basic chemical and crystalline structure, but rather, due to the interaction of light with certain inclusions or structural features within the gem. Major Optical Phenomena in Gemstones: labradorescence, play of color, adularescence, aventurization, chatoyancy, asterism, color change.

Labradorescence is a type of iridescence caused by repeated, microscopically thin layer (lamellar) twinning in labradorite feldspar.

Iridescence in precious opal is correctly called "play of color" or "color play". What is taking place in opals is not dispersion, but iridescence. We divide all opals (a huge group of gems) into precious and common, based on whether they have color play, or not, respectively.

In moonstone, adularescence is due to a layer effect, where thin inner strata of two

types of feldspar intermix. These layers scatter light either equally in all spectral regions producing a white demitint, or as in the most valuable specimens, preferentially in the blue or the blue and orange. As in so many cases of optical phenomena the size or distance from layer to layer influences the colors we see.

Aventurization is a consequence of reflection. When disk or plate-like inclusions of another mineral are present, and are of a highly reflective nature such that they act as tiny mirrors, the gem sparkles and glitters. This glitter is called aventurization. The most common reflectors are copper, hematite and mica.

Chatoyancy is also due to reflection, but in this case, rather than involving plate-like inclusions scattered randomly, it is due to parallel thread-like reflective inclusions such as needles or tubes.

Asterism is essentially a special case of the cat's eye effect, where the inclusions responsible for reflections are oriented parallel to more than one axis in the crystal. As with cat's eyes, the stone must be both properly oriented, and cut in a high dome to display the star.

A color change gem is one whose color is substantially different when viewed with an incandescent light source as compared to its color as seen under daylight or a daylight equivalent fluorescent source. Due to this phenomenon's strong association with the alexandrite variety of chrysoberyl, it is sometimes termed the "alexandrite effect", regardless of which species is displaying it.

Part 7 Text

Major Optical Properties of Gems

Color

The color of a gem is determined by selective absorption of some of the wavelengths of light. We know that what appears to us as white (or colorless) light is actually made up of light of various colors.

There are three aspects to a formal colored-stone color description: hue, tone, and saturation. Using these three descriptors, very detailed and nuanced color discriminations can be made, and communicated, between gemologists, jewelers and gem buyers. Let's take a look at them each in turn.

Hue: the hue of a gem is its basic position in the color spectrum, including red, orange, yellow, green, blue or violet, but it also includes all the possible intermediates like slightly yellowish orange, or moderately bluish green.

Tone: the tone of a gem, basically how light or dark the color, is independent of its hue and ranges from so light as to appear virtually colorless, to so dark as to look black.

Saturation: the least commonly quantified aspect of gem color is "saturation", which is a measure of the purity of color, that is, the relative presence or absence of modifying grey or brown hues. It turns out that in most cases, as long as the hue and tone are reasonably nice, the degree of saturation of color is the prime value factor in gemstones.

Luster

The luster of a gemstone is comprised of the quantity and quality of the light reflected from its surface. There is an inherent, potential luster possible for each species and variety of gemstone. The actual luster, on any individual piece; however, may be less than this, due to the skill level of the lapidary, the facet condition, the presence of inclusions, or various chemical or physical changes, such as oxidation or abrasion, that can affect the surface.

gemstones	luster	gemstones	luster	gemstones	luster
pyrite	metallic	diamond	adamantine	zircon	sub-adamantine
fire agate	vitreous	fluorite	sub-vitreous	nephrite	greasy
amber	resinous	pearl	pearly	tiger's eye	silky

The names, which have been given to the various lusters seen in gems, are derived from their resemblance to familiar surfaces(The prefix "sub" indicates "just less than"). Some lusters are so embodied by a particular stone, that its appearance is named for that stone, as in the case of adamantine luster (adamas- Greek for diamond), and pearly luster. Looking through either of your textbooks at the descriptions of the various gems will convince you that a substantial majority of gems have a glass-like or "vitreous" luster.

Transparency

The degree of transparency of a gemstone is one of its most directly observable and familiar characteristics.

Transparency (or lack of it) is dependent on how much light gets through the gem, and is affected not only by the chemical and crystalline nature of the gem, but also by its thickness and, as in the case of luster, by inclusions, and its surface condition. In the discussion and examples that follow below, we will be looking at the "potential" maximum transparency of a species in general, rather than the actual transparency of any individual specimen.

When light hits the surface of a gem, there are only three fates for it (with respect to transparency). Various portions of the total amount of light will be reflected, absorbed or transmitted. The proportion in each category will determine the transparency of that gem.

Reflection

Light is reflected when it hits an exterior or interior surface of the gem and is bounced back off, or out of, the gem, in the direction of the observer.

Dispersion

Dispersion (sometimes called "fire"), is the separation of white light into its spectral colors. It may be observed as specks of red, blue or green which flicker as the gem is turned.

The exact figures for dispersion can be painstakingly measured in a laboratory setting using special equipment, and then calculated as the difference between the RIs of red light and violet light in a given species. Potential dispersion in gems, thusly measured, ranges from 0.007 to 0.280.

Outside the lab, dispersion is generally judged visually, without instruments, simply as: absent, slight, moderate, strong or very strong. The degree of visible dispersion is affected by species (due to RI), but also by the body color, and cut proportions of the gem.

Examples of gems with slight dispersion potential are: fluorite (0.007), common glass ("crown" or silica glass) (0.010) and quartz (0.013). Regardless of color or cut, these gems just aren't going to show visible dispersion, the effect is too slight.

Those with moderate potential for dispersion include: tourmaline (0.017), corundum (0.018) and spinel (0.020). Such gems rarely show visible dispersion, but an occasional light colored specimen of substantial size with very high crown angles may do so.

Examples of strongly dispersive gems are: zircon (0.038), diamond and benitoite (both 0.044), and cubic zirconia (a synthetic) (0.066). Gems in this range will usually show dispersion. Exceptions might be those of very dark body color, or small pieces cut with rather low crown angles.

Very strongly dispersive gems include: sphalerite (0.156), strontium titanate (a synthetic) (0.190) and synthetic rutile (0.280). There would be very few cases where a gem in this group did not show substantial dispersion.

Diamond is the most well known gem that shows dispersion, and it is one of that gem's most appealing attributes. The success of either a natural or man-made diamond simulant, depends to a large extent on how well the substitute matches diamond in this characteristic.

Refraction

SR stands for singly refractive. In such gems, each beam of light entering the gem stays as a single beam which has a single refractive index (travels at the same speed), regardless of the direction from which it enters. In this group we find all amorphous gem materials, such as opal, glass, amber, etc. as well as all crystalline gems belonging to the cubic (isometric) system. The most commonly encountered gems of the cubic system are: diamond, garnet and spinel.

DR stands for doubly refractive. In such gems, single beams of light upon entering the gem, are split into two separate beams, which then travel perpendicularly to each other. Each of the resultant beams takes a different path through the crystal and, consequently, has its own speed. Such gems, then, have two RIs. In this group are all the gems of the non-cubic crystal systems.

Birefringence, a property of DR gems only, is measured as the difference between the high and low RIs of the split beams. It ranges from a low of 0.003 to a high of 0.287.

When a transparent gem with high DR is faceted, and the view through the table direction of that gem is not in an optic axis direction, the slightly "out of sync" light beams can create an appearance of interior "fuzziness" or in larger stones, can show up as two distinct images of each facet edge. This is known as "facet doubling" and it can be a pain in the neck to a facetor who, in trying to prevent it, must find an optic axis direction for the table of the stone. But it can also be a valuable identifying characteristic that can often be seen with the naked eye or a simple 10×loupe.

Pleochroism

Pleochroism is the property of DR gems which results in their showing different colors, or different shades of the same color, when viewed in different crystal axis directions.

This effect can be weak, moderate or strong, depending on the species of gem, the colors involved, and also on the color tone of the particular piece. A very light piece of a pleochroic species will show the effect less clearly than a more richly colored one. Unless the

effect is extreme (as it is in iolite and andalusite), we generally do not see it in a cut gem, because the bouncing and mixing of the light caused by the internal reflections from facets and edges blends the colors together and obscures it.

Dichroic gems (like corundum) show two different colors while trichroic gems (like iolite) show three.

Pleochroism will not be observed in SR gems, nor in DR gems when looking through an optic axis direction.

Fluorescence

When a gem absorbs either SW or LW UV, or both, and immediately emits visible light, the phenomenon is called fluorescence.

Fluorescence can be absent, in which case we say the gem is inert, or present in weak, moderate or strong form. The light emitted by fluorescence can be the same color or a different one from the color of the gem itself, and a gem can have the same or different reactions to LW and SW.

Natural, white to light yellow diamonds often fluoresce blue, about 30% of them do so.

Part 8　Words and Expressions

optical[ˈɒptɪkl]	adj. 光学的
luster[ˈlʌstə]	n. 光彩,光泽
dispersion[dɪsˈpɜːʃn]	n. 散射
pleochroism[pliˈɒkrəʊɪzəm]	n. 多色性
chatoyancy[ʃəˈtɔɪənsi]	n. 猫眼效应
asterism[ˈæstərɪzəm]	n. 星光效应
chrysoberyl[ˈkrɪsəberɪl]	n. 金绿宝石
alexandrite[ˌælɪɡˈzɑːndraɪt]	n. 变石
iridescent[ˌɪrɪˈdesnt]	adj. 有晕彩的,彩虹色的,闪光的
amazonite[ˈæməzənaɪt]	n. 天河石
mount[maʊnt]	vt. 镶嵌
chicken-blood stone	鸡血石
spinel[ˈspɪnəl]	n. 尖晶石
irradiate[ɪˈreɪdieɪt]	v. 照射
assessment[əˈsesmənt]	n. (为征税对财产所作的)估价,被估定的金额
fluorite[ˈflʊəraɪt]	n. 萤石
moissanite[ˈmɔɪsænaɪt]	n. 碳硅石,莫桑石
appreciation[əˌpriːʃiˈeɪʃn]	n. 欣赏,正确评价
refraction[rɪˈfrækʃn]	n. 折光,折射
fire[ˈfaɪə]	n. 火彩
prismatic[prɪzˈmætɪk]	adj. 棱镜的
dichroism[ˈdaɪkrəʊɪzəm]	n. 二色性
reflected[rɪˈflektɪd]	adj. 反射的
metallic[məˈtælɪk]	adj. 金属(性)的
adamantine[ˌædəˈmæntaɪn]	adj. 金刚石似的
vitreous[ˈvɪtriəs]	adj. 玻璃质的
resinous[ˈrezɪnəs]	adj. 树脂质的
greasy[ˈɡriːsɪ]	adj. 油脂的
silky[ˈsɪlki]	adj. 丝的,柔滑的
pearly[ˈpɜːli]	adj. 珍珠似的
dull[dʌl]	adj. 阴暗的,感觉或理解迟钝的
	vt. 使迟钝,使阴暗,缓和
encompass[ɪnˈkʌmpəs]	v. 包围,环绕
crystalline[ˈkrɪstəlaɪn]	adj. 晶质的
interaction[ˌɪntərˈækʃn]	n. 交互作用
inclusion[ɪnˈkluːʒn]	n. 包裹体,内含物
labradorescence[ˌlæbrədɔːˈresəns]	n. 拉长晕彩,变彩

play of color	变彩
adularescence[ˌædʒuləˈresns]	冰长石光彩，月光效应
microscopically[ˌmaɪkrəuˈskɔpɪkəlɪ]	adv. 显微镜地，精微地
feldspar[ˈfeldspɑː]	n. 长石
respectively[rɪˈspektɪvlɪ]	adv. 分别地，各个地
moonstone[ˈmuːnstəun]	n. 月光石
strata（stratum 的复数）	n. 地层
scatter[ˈskætə]	v. 分散
demitint[ˈdemɪtɪnt]	n. 晕色
preferentially[prefəˈrenʃəlɪ]	adv. 先取地，优先地
sparkless[ˈspɑːklɪs]	adj. 不发出闪光的
glitter[ˈglɪtə]	n. 闪光
hematite[ˈhemətaɪt]	n. 赤铁矿
mica[ˈmaɪkə]	n. 云母
needle[ˈniːdl]	n. 针
tube[ˈtjuːb]	n. 管，管子
responsible[rɪsˈpɔnsəbl]	adj. 有责任的，可靠的，可依赖的，负责的
oriented[ˈɔːrɪentɪd]	adj. 定向的
dome[dəum]	n. 圆屋顶
substantially[səbˈstænʃəlɪ]	adv. 充分地
incandescent[ɪnkænˈdesnt]	adj. 白炽的，遇热发光的
fluorescent[fləˈresənt]	adj. 荧光的
determine[dɪˈtəːmɪn]	v. 决定，确定
hue[hjuː]	n. 色彩，颜色
tone[təun]	n. 色调　　vi. 颜色调和
saturation[sætʃəˈreɪʃən]	n. 饱和度
nuanced[ˈnjuːɑːnst]	（色彩、音调、措词、意味、感情等）有细微差别的
discrimination[dɪsˌkrɪmɪˈneɪʃn]	n. 辨别，区别，识别力，辨别力，歧视
communicate[kəˈmjuːnɪkeɪt]	v. 沟通，通信，（房间、道路、花园等）相通，传达，感染
spectrum[ˈspektrəm]	n. 光谱
violet[ˈvaɪələt]	n. 紫罗兰
intermediate[ˌɪntəˈmiːdɪət]	adj. 中间的
quantify[ˈkwɔntɪfaɪ]	vt. 确定数量　v. 量化
purity[ˈpjuərɪtɪ]	n. 纯度
prime[praɪm]	adj. 主要的，最初的，最好的，第一流的，根本的
inherent[ɪnˈhɪərənt]	adj. 固有的，内在的
lapidary[ˈlæpɪdərɪ]	n. 宝石商
oxidation[ɔksɪˈdeɪʃn]	n. 氧化
abrasion[əˈbreɪʒn]	n. 磨损

Unit 2 Optical Properties of Cut Gemstones

zircon['zɜːkɒn]	n. 锆石
fire agate	n. 火玛瑙
tiger's-eye	n. 虎睛石
Fates[feɪts]	n.〈希神〉命运三女神
category['kætɪɡərɪ]	n. 种类,别
flicker['flɪkə]	n. 闪烁,闪光,扑动,颤动 vi. 闪动,闪烁
painstakingly['peɪnsˌteɪkɪŋlɪ]	adj. 值得花力气的,值得下功夫的
thusly['ðʌslɪ]	adv. 因而,从而,这样,如此
crown[kraʊn]	n. 王冠,花冠,顶
	vt. 加冕,顶上有,表彰,使圆满完成
corundum[kəˈrʌndəm]	n. 刚玉,金刚砂
benitoite[bəˈniːtəʊaɪt]	n. 蓝锥矿
cubic zirconia['kjuːbɪk][zɜːˈkəʊnɪə]	n. 立方氧化锆
sphalerite['sfælərait]	n. 闪锌矿
strontium titanate['strɒnʃɪəm]['taɪteɪneɪt]	n. 钛酸锶
rutile['ruːtiːl]	n. 金红石
reflective index	n. 折射率
amorphous[əˈmɔːfəs]	adj. 非晶质的
cubic system	n.(等轴)晶系
perpendicularly[ˌpɜːpənˈdɪkjʊləlɪ]	adv. 垂直地,正交地
sync[sɪŋk]	n. 同时,同步
fuzziness[ˈfʌzɪnɪs]	n. 绒毛,模糊,失真的
facet doubling	刻面棱重影
loupe[luːp]	n. 小型放大镜
iolite['aɪəlaɪt]	n. 堇青石
andalusite[ˌændəˈluːsaɪt]	n. 红柱石
dichroic[daɪˈkrəʊɪk]	adj. 二色性的
trichroic[traɪˈkrəʊɪk]	adj. 三色的
fluorescence[fləˈresəns]	n. 荧光
inert[ɪˈnɜːt]	adj. 惰性的
emit[ɪˈmɪt]	vt. 发出,放射
reaction[rɪˈækʃn]	n. 反应

Part 9 Check Your Understanding

Exercise 1 Answer the following questions

(1) Illustrate five optical properites of gems.
(2) How do you describe the color of gemstone?
(3) Please describe the luster of fluorite, ruby and amber?
(4) Which factor can impact transparency of gemstones?

(5) Illustrate three strongly dispersive gems.

(6) What is the pleochroism of gemstone?

(7) Which gem can appear "facet doubling"?

Exercise 2　Fill in the blanks

(1) The luster of gems is characterized by the terms ＿＿＿＿, ＿＿＿＿, ＿＿＿＿, ＿＿＿＿, ＿＿＿＿, ＿＿＿＿.

(2) Illustrate major optical phenomena in Gemstones: ＿＿＿＿, ＿＿＿＿, ＿＿＿＿, ＿＿＿＿, ＿＿＿＿, ＿＿＿＿.

(3) ＿＿＿＿ can appear play of color.

(4) Aventurization is a consequence of ＿＿＿＿.

(5) The most common reflectors in Aventurization Phenomenon are ＿＿＿＿, ＿＿＿＿, ＿＿＿＿.

Exercise 3　Translation

(1) The beauty of gems depends to a large extent on their optical properties. The most important optical properties are the degree of refraction and color. Other properties include fire, the display of prismatic colors; dichroism, the ability of some gemstones to present two different colors when viewed in different directions; and transparency.

(2) Chatoyancy is also due to reflection, but in this case, rather than involving plate-like inclusions scattered randomly, it is due to parallel thread-like reflective inclusions such as needles or tubes.

(3) Fluorescence can be absent, in which case we say the gem is inert, or present in weak, moderate or strong form.

(4) 欧泊可具有变彩效应，玻璃可具有猫眼效应，金绿宝石可具有变色效应。

(5) 有的宝石是单折射，有的宝石是双折射，具双折射的宝石可能见到多色性。

(6) 钻石和合成立方氧化锆都具有较高的色散。

Unit 3　Physical Properties of Gems

Part 1　Dialogue

本章音频

Grace: Would you like jewelry?
Kitty: Yes, of course.
Grace: What kind of jewelry do you like to have?
Kitty: I'd like 14K and 18K gold necklaces, chains and earrings.
Grace: Sure. Here is a nice gold necklace.
Kitty: Can I have a look?
Grace: Here you are. Its regular price is $5600, and now you can have it with a ten percent discount.
Kitty: It's very elegant. I'll take it. Besides, I should like to look at some bracelets.

Grace: Jadeite or nephrite?
Kitty: I don't know what you said, but I love the bracelets with light purple.
Grace: Oh, that's jadeite from Burma.
Kitty: What's the price for it?
Grace: $1500.
Kitty: How about nine hundred dollars?
Grace: I'm sorry we only sell at fixed prices.

Kitty: It's a pity, I'm ready to give it up. Food is more important than jewelry, I think.
Grace: Haha, I see. This is the bill of the necklace, the counter is on the right.
Kitty: Yes, I am going to pay the check, bye-bye.

Part 2　Fill the Chinese Meanings with the Teacher's Tutorship

cleavage _____　　fracture _____　　peridot _____　　specific gravity _____
hardness _____　　moissanite _____　　fluorite _____　　carat _____
tenacity _____　　toughness _____　　feldspar _____　　piezoelectricity _____

Part 3　Link the Words and the Relevant Pictures

fracture
cleavage
conchoidal
hardness:7
doubly
refractive
fluorite
thermoele-
ctricity

Part 4　Useful Phrases

China jade artworks show and award　中国玉雕作品展示评奖活动
fair for investment and trade in the jewelry industry　珠宝项目投资贸易洽谈会
sees and hears about Burma Gems　缅甸宝石见闻
jewelry moulding　首饰造型
recommendations for jewelry wearers　首饰选配建议
handicraft fossil　工艺化石
how to become an excellent appraiser　如何成为更优秀的鉴定人员
masterpieces of jadeite　翡翠杰作
Tanzania Gems　坦桑尼亚宝石
application of new technology to gemology　新技术在宝石学上的应用
snuff bottle　鼻烟壶
hair ornaments of empress in Qing Dynasty　清朝皇后头饰
semifinished diamond marketing　钻石毛坯的销售
some hints for jewelry jade-ware shoppers　如何引导顾客选购珠宝
China jewelry auction　中国珠宝拍卖

Part 5 Some Sentences

(1) A distinctive, common fracture is called conchoidal, which is a shell-like break. This break is seen in glass, quartz, opal, peridot and amber, to name a few. Other possible fractures include uneven, splintery, granular, or subconchoidal.

(2) A knowledge of cleavage for the cutter is important as it can lead to an easy first step to the fashioning process for diamonds. When considering colored stones, cleavage is avoided as it is very difficult to polish a gem parallel to a cleavage plane.

(3) Although SG measurements can be made on either rough or cut gems, the gems must be unmounted, and composed of a single material. You cannot do a SG measurement on a gem that is set in a piece of jewelry, or on an assembled stone, like a doublet. Porous gems cannot be measured, as the liquid they absorb affects the SG measurement and in some cases, can harm the stone.

(4) A fair or poor tenacity does not mean the gem is less valuable, but does have implications for care and cleaning as well as setting the stone in a secure, protective mounting.

Part 6 Short Paragraph

Weighing Gems

In the early history of gem marketing, depending on the geographic location, one of two common items, familiar to both buyers and sellers, was used to measure the amount of gem material being bought and sold: the wheat grain and the carob seed. Each of these commodities was known for being particularly uniform in size and weight. We still see remnants of this early system in today's terms: "carat" the international metric unit used for gems, and "grain" a unit sometimes used in selling pearls, and also in today's system of apothecary measure.

The carat, pronounced like the vegetable "carrot" and abbreviated "ct" is 0.2 grams. So, there are five carats per gram. The metric system is the basic international standard used for gem commerce. Many of us who live in the US or UK where English measure is more common, need to take time, and do some practice, in order to get a "feel" for carats, grams,

etc. The ounce, a familiar English unit of weight, equals approximately 142 carats. So, there really isn't an appropriately small unit in the English system which could be easily applied to gem weights.

Another oddity of the US system is our use of the term "karat", also pronounced like the vegetable carrot, but abbreviated "K" or "Kt" to indicate the fineness(purity) of gold. In most other countries, the purity of gold is indicated by the number of parts of gold out of 1000, such as 585 or 750, so there is no chance of confusion with gem weights. The number 585 means that 585 out of 1000 parts of the alloy are gold or, in other words, that the gold content is 58.5%.

In comparison, the karat system uses the number of parts out of 24 that are gold. 24K means 24/24th, pure gold, also known as "fine" gold, 18K gold=18/24th gold, and 14K=14/24th gold. (14K and 18K and 24K translate then, in the international system, to 585, 750 and 999 respectively).

Part 7 Text

Physical Properties of Gems

The mineral's composition and crystalline structure impart the various physical proper-

ties that characterize each specimen. Knowledge of the properties of gemstones is important for the gem cutter and setter, as well as to the consumer who can use that information to care for the gem.

Specific Gravity

Specific gravity, also known as relative density, differs widely among gemstones, and is one of their most important physical characteristics from the viewpoint of gem identification. Specific gravity (SG) is the ratio of the weight of one unit volume of the gem to the weight of the same unit of water. For example, to say sapphire (corundum) has SG=4.0, means precisely that a cubic inch of sapphire weighs four times as much as a cubic inch of water. In natural gems, SG values range from just over 1 (1.08 for amber) to just short of 7 (6.95 for cassiterite).

Unit 3　Physical Properties of Gems · 27 ·

There are several ways to directly measure the specific gravity. To arrive at a relative measure of specific gravity, heavy liquids are used. By far the most precise technique for SG determination involves use of a specially modified weighing balance that allows a gem sample to be weighed in air (Wa), and also weighed in water (Ww). Using Archimedes Principle: "A body immersed in water weighs less by the volume of water displaced", and a simple calculation, SG can be determined with substantial accuracy.

Hardness

The hardness of the mineral refers to its resistance to scratching and abrasion and also to the cutting resistance. The more resistant the surface is to scratching, the harder the mineral is, and the stronger the bonding forces are holding the atoms together. Gemstones are often tested by using the Mohs' hardness scale to determine just how hard they are.

1.滑石；2.石膏；3.方解石；4.萤石；
5.磷灰石；6.正长石；7.石英；8.黄玉；
9.刚玉；10.金刚石

Hardness	1	2	3	4	5
Mineral	Talc	Gypsum	Calcite	Fluorite	Apatite
Hardness	6	7	8	9	10
Mineral	Feldspar	Quartz	Topaz	Sapphire	Diamond

You can see that topaz has a hardness of 8 on the scale. This means that a topaz is harder than anything except a sapphire or a diamond.

Cleavage and Fracture

Cleavage and fracture refer to the characteristic manner in which gems will break when

an external force or stress is applied. Some minerals have a special way of breaking parallel along planes of atomic weakness, creating smooth flat surfaces. This break is called cleavage. Crystalline minerals have cleavage and fracture, whereas amorphous or massive stones only fracture.

In rough material, a cleavage break may already be obvious or it can be determined by giving the specimen a tap with a hammer. Rough diamond is often cleaved and then cut into shapes. Cleavage is not possible to observe in fashioned gems unless an internal imperfection can be observed or there is an accidental blow struck along a cleavage direction and the gem breaks. Thus, diamond has very well developed cleavage and although it is the hardest known substance, the ready cleavage makes it suspectable to damage.

Fracture is the way a stone breaks. It is a break in a direction other than along cleavage planes and results when the bonding forces are similar in all directions. Consider fracture to be similar to a piece of wood breaking in a direction other than the direction of it's grain.

Tenacity or Toughness

Tenacity or toughness is the ability of a stone to withstand pressure or impact. It is the resistance to crushing, breaking, or tearing. Minerals which crumble into small pieces or a powder are said to be brittle.

Piezoelectricity

Piezoelectricity, or pressure electricity, is found in minerals that have polar axes or lack a center of crystalline symmetry. The crystal axes have different properties at the opposite ends of the polar axis and when pressure is exerted at these ends, electricity can flow creating opposite positive and negative ends. Quartz and tourmaline are piezoelectric.

Thermal Conductivity

Some stones are good conductors of heat, such as quartz, which draws heat away from the body when held and thus feels cold to the touch. Heat is conducted differently in various minerals according to their crystal system. A poor thermal conductor, such as amber, feels warm to the touch because it

does not conduct heat away from the body. The surface of a genuine gemstone will conduct heat more rapidly than that of glass or an artificial stone.

Thermal conductivity should also be considered when cutting gemstones, as some stones will need a cooling-off period during the cutting. This is also used in thermal conductivity instruments to differentiate diamond which conducts heat very well from its simulants and imitations. Some instruments use it to identify other gemstones, but they are expensive and of value only when used with care and some gemological knowledge.

Part 8 Words and Expressions

chain[tʃeɪn]	n. 链（条）
discount[ˈdɪskaʊnt]	n. 折扣
elegant[ˈelɪɡənt]	adj. 文雅的，端庄的，雅致的，〈口〉上品的，第一流的
cleavage[ˈkliːvɪdʒ]	n. 解理
fracture[ˈfræktʃə]	n. 断口，裂隙
peridot[ˈperɪdɔt]	n. 橄榄石
density[ˈdensɪtɪ]	n. 密度
specific gravity[spɪˈsɪfɪk][ˈɡrævəti]	n. 比重
tenacity[tɪˈnæsəti]	n. 坚韧
piezoelectricity[paɪiːzəʊlekˈtrɪsɪtɪ]	n. 压电（现象）
conchoidal[kɔŋˈkɔɪdl]	adj. 贝壳状的
thermoelectricity[θəːməʊlekˈtrɪsɪtɪ]	n. 热电，温差电
Burma[ˈbɜːmə]	n. 缅甸（东南亚国家）
mould[məʊld]	n. ［亦作 mold］模具
handicraft[ˈhændɪkrɑːft]	n. 手工艺品
fossil[ˈfɔsl]	n. 化石
masterpiece[ˈmɑːstəpiːs]	n. 杰作
Tanzania[ˌtænzəˈniːə]	n. 坦桑尼亚（东非国家）
application[ˌæplɪˈkeɪʃn]	n. 申请
semifinished[ˈsemɪfɪnɪʃt]	adj. 半完成的
splintery[ˈsplɪntəri]	adj. 裂片似的，易碎裂的，碎裂的
granular[ˈɡrænjʊlə]	adj. 粒状的
parallel[ˈpærəlel]	adj. 平行的
rough[rʌf]	adj. 粗糙的，粗略的
assemble[əˈsembl]	vi. 集合

assembled	装配,组合
doublet['dʌblɪt]	n. 二层石
porous['pɔːrəs]	adj. 多孔渗水的
geographic[ˌdʒɪəˈɡræfɪk]	adj. 地理学的,地理的
grain[ɡreɪn]	n. 谷物,谷类,谷粒,细粒,颗粒,粮食
carob['kærəb]	n. 长豆角
metric['metrɪk]	adj. 米制的,公制的
apothecary[əˈpɒθəkərɪ]	n. 药剂师,药师
abbreviated[əˈbriːvieɪtɪd]	adj. 简短的
gram[ɡræm]	n. 克
ounce[auns]	n. 盎司
approximately[əˈprɒksɪmətlɪ]	adv. 近似地,大约
oddity['ɒdɪtɪ]	n. 奇异,古怪,怪癖
confusion[kənˈfjuːʒən]	n. 混乱,混淆
viewpoint['vjuːpɔɪnt]	n. 观点
ratio['reɪʃɪəu]	n. 比,比率,[财政]复本位制中金银的法定比价
cassiterite[kəˈsɪtəraɪt]	n. 锡石
Archimedes[ˌɑːkɪˈmiːdiːz]	n. 阿基米德
principle['prɪnsəpl]	n. 法则,原则,原理
immersed[ɪˈmɜːst]	adj. 浸入的
resistance[rɪˈzɪstəns]	n. 反抗,抵抗,抵抗力,阻力,电阻,阻抗
abrasion[əˈbreɪʒn]	n. 磨损
bond[bɒnd]	v. 结合
atom['ætəm]	n. 原子
mineral['mɪnərəl]	n. 矿物,矿石
talc[tælk]	n. 滑石
gypsum[ˈdʒɪpsəm]	n. 石膏
calcite['kælsaɪt]	n. 方解石
apatite['æpətaɪt]	n. 磷灰石
manner['mænə]	n. 方式
massive['mæsɪv]	adj. 大块的
tap[tæp]	n. 轻打,活栓,水龙头
suspectable[səsˈpektəbl]	adj. 可疑的
withstand[wɪðˈstænd]	vt. 抵挡,经受住
crumble[ˈkrʌmbl]	v. 弄碎,粉碎,崩溃
brittle['brɪtl]	adj. 易碎的
polar['pəulə]	adj. 极性的
axes['æksiːz]	n. 轴线,轴心
positive['pɒzətɪv]	adj. 正的,阳的

negative[ˈneɡətɪv]	n. 否定,负数,底片
	adj. 否定的,消极的,负的,阴性的
thermal conductivity	导热性
thermal conductor	热导体
genuine[ˈdʒenjuɪn]	adj. 真实的,真正的,诚恳的
demist[ˌdiːˈmɪst]	vt. 为……除雾
cooling-off period	n. 冷却期
differentiate[ˌdɪfəˈrenʃieɪt]	v. 区别,区分

Part 9 Check Your Understanding

Exercise 1   Answer the following questions

(1) How do you describe the fracture appearance of gemstones?
(2) How does the cleavage work on the cut of gemstones?
(3) What do we pay attention to when measuring the SG?
(4) How do you explain the term "carat"?
(5) Please give an example to show the method of measuring the SG.

Exercise 2   Fill in the blanks

(1) 1 carat is equal to _____ gram.
(2) Please illustrate several physical properties of gems: ____, ____ and ____.
(3) Please illustrate the ten normative Mohs' hardness of minerals: _____, _____, _____, _____, _____, _____, _____, _____, _____, _____.
(4) Cleavage and fracture refer to the characteristic manner in which gems will _____ when an external force or stress is applied.
(5) _____ is possibly provided with piezoelectricity.

Exercise 3   Translation

(1) The carat, pronounced like the vegetable "carrot" and abbreviated "ct" is 0.2 grams. So, there are five carats per gram. The metric system is the basic international standard used for gem commerce.
(2) Specific gravity, also known as relative density, differs widely among gemstones, and is one of their most important physical characteristics from the viewpoint of gem identification.
(3) The hardness of the mineral refers to its resistance to scratching and abrasion and also to the cutting resistance.
(4) 红宝石的标准摩氏硬度为9。
(5) 敲击萤石将会看见较光滑的解理面。
(6) 可以利用钻石较高的导热性鉴别出钻石的仿制品。

Unit 4 Gemological Instruments

Part 1 Dialogue

Mary: Are you looking for something, Sir?
Tom: I want to buy some jewelry for my girl friend.
Mary: What kind of jewelry do you like?
Tom: Oh, I have no idea.
Mary: Why not look at these pretty brooches?
Tom: Pure gold or carats?
Mary: Karat gold ones.
Tom: Give me one with an emerald gemstone.
Mary: Certainly.
Tom: What's the price for it?
Mary: Nine hundred and fifty dollars.
Tom: I am wondering whether you could give me a discount?
Mary: I'm sorry we only sell at fixed prices.
Tom: OK. I'll make a decision after comparing the prices of the similar.
Mary: Oh, that is a good idea.
Tom: Thank you, Bye.
Mary: Bye-bye.

本章音频

Part 2 Fill the Chinese Meanings with the Teacher's Tutorship

microscope _____   10×triplet loupe _____   phosphorescence/fluorescence _____
singly refractive _____   doubly refractive _____   refractive index _____
magnification _____   lens _____   polariscope _____   luminescence _____
spectroscope _____   dichroscope _____   pleochroism _____   filter _____

Part 3 Link the Words and the Relevant Pictures

loupe
microscope
refractometer
polariscope
fluorometry
dichroscope

Part 4 Useful Phrases

wedding diamond ring marketing   结婚钻戒销售
jewelry designer   首饰设计师
mini diamond market   碎钻市场
colourless synthetic diamond   无色合成钻石
diamond polishing trade now and look ahead   钻石加工业现状及展望
gem refractometer theory   宝石折射仪原理
vice-president of sales   销售副总裁
senior customer manager   高级客户经理
sales manager   销售经理
regional sales manager   地区销售经理
merchandising manager   采购经理
sales assistant   销售助理
buy two get one free   买二赠一
customer care is our top priority   顾客至上
peace of mind from the minute you buy   买着放心

Part 5 Some Sentences

(1) When white light interacts with a gemstone, a portion is reflected off the surface, while another portion, that enters the stone, is slowed down and "bent" or refracted. This optical phenomenon gives rise to an important measurable constant termed the refractive index.

(2) Bench type polariscopes are used to quickly and easily check if gemstones are single or double refracting.

(3) Prism type spectroscope.

Uses: ① Unpolished stones.

② To identify treated stones.

③ Faceted stones that have a refractive index above the normal range of the refractometer.

④ Identify some synthetics (i.e. Natural Blue Sapphire from its synthetic counterpart).

Disadvantages: Wavelengths are not linearly spaced out. The red end is bunched while the blue/violet is spread out.

(4) Diffraction grating spectroscope.

Uses: Same as the prism type spectroscope.

Disadvantages: Spectrum is not as bright. Hard to regulate the amount of light that enters the instrument. Hard to view in the blue end of the spectrum.

Advantages: Cost, they are relatively inexpensive.

Part 6 Short Paragraph

Different mensuration of gem's refractive index

A different method used to determine the refractive index is with immersion. Immersing gem material in a liquid with approximately the same refractive index, will make the gem nearly invisible. Immersing the gem in a liquid with a different refractive index can make the gem stand out clearly or in high relief. This method is used with unknown minerals and requires crushing the specimen to view the fragments in different refractive index liquids with the aid of a microscope. Obviously this is not the test for cut and fashioned stones and requires a set of refractive index liquids and a high-powered microscope.

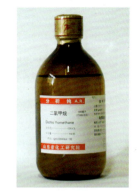

The immersion method is sometimes used without powdering the gem, and simply immersing it in different liquids in an "immersion cell" or small dish. The dish is suspended over a white paper

that would allow for a black card to be passed beneath it; when observing from above, if the edge of the card seen through the liquid and stone is a straight line the refractive index of the liquid and stone are the same. If the stone has a lower index, the card's edge seems to move ahead of the edge seen through the adjacent liquid or just the opposite if the stone has a higher index.

Some common immersion liquids include: water (1.33), ethyl alcohol (1.36), acetone (1.36), glycerine (1.46), olive oil (1.48), xylene (1.49), clove oil (1.53), ethylene bromide (1.54), bromoform (1.60), methylene iodide (1.74). Relief can sometimes be observed when using the heavy liquids for specific gravity testing. When a stone is dropped into the liquid to observe if it floats or sinks, it sometimes vanishes! That is the liquid and stone are of a similar refractive index.

Part 7   Text

Gemological Instruments

10× Triplet Loupe and Microscope for Testing Magnified Observations

The function of magnification, whether by a hand held lens or the microscope, is to enlarge an image of the object so details become clear on both the surface and interior. The hand lens or loupe is a single lens system or simple microscope, with a variety of magnifications possible. The triple aplanatic is a high quality lens made of two external lenses of flint glass (or lead glass). [Crown

glass is common bottle glass made of silica, soda and lime; while flint or lead glass, the glass used to imitate gems, is composed of silica, soda and a lead oxide.] Although the hand lens comes in different magnifications, the 10× (10 power) magnification is good for most gem purposes. A magnification higher

than 10× creates difficulty in illuminating the stone, and reduces the field of view and depth of field.

For a thorough examination, the binocular stereoscopic microscope is a versatile and very effective instrument to view gemstones. Besides the ocular or eyepiece lenses, there is an objective lens, which creates a two lens system to produce a enlarged sharp image.

The microscope magnification is calculated by multiplying the objective and ocular magnification (e. g., a 10× objective and 3× ocular produces a magnification of 30 times). Gem microscopes create a "reinversed" image (regular compound microscopes invert the image) and usually have a zoom feature to vary the magnification continuously. The overhead lighting source is fluorescent, while a high intensity light is transmitted up from the base and through the stone. The light in the base is on a rheostat for intensity control and fitted over with an iris diaphragm, which opens and closes controlling the amount of light that can enter the stone from below.

Polariscope for Testing Refraction

Gemstones can be divided into two groups, isotropic or singly refractive and anisotropic or doubly refractive. Isotropic or singly refractive gemstones include minerals in the isometric crystal system and amorphous material, such as glass, opal, garnet and diamond. These gemstones have one refractive index, that is the light that plane polarized, it moves through the stone with an equal velocity in all directions. Minerals falling into any other crystal system (tetragonal, orthorhombic, hexagonal, monoclinic and triclinic) are doubly refractive or anisotropic. These gemstones have two or

three refractive indices, that is the light is split into different directions with the velocity varying with crystallographic axis directions. When light enters a doubly refractive stone or anisotropic crystal, it is separated into two polarized rays vibrating in mutually perpendicular planes. The rays travel at different velocities and two indices of refraction can be measured on a refractometer and detected with a polariscope. The polariscope is the gem instrument used to separate the two groups, using two filters, an analyzer and polarizer.

Polariscope is to make quick determination of single or double refraction. But most gemologists don't use it to its fullest extent, that being with a conoscope to determine optic character. But a nice instrument to have for SR or DR stones. But let's face it, the most use these really get is for the refractometer.

Dichroscope for Testing Pleochroism

Light that passes through a doubly refractive gemstone or anisotropic mineral is split in different directions with unequal velocity. The light is absorbed differently in different vibration directions, resulting in color variation known as pleochroism. In minerals with only two rays, two pleochroic colors can be detected, called dichroism. In minerals with three principal vibration directions (three refractive indices), three different pleochroic colors can be

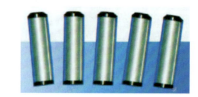

detected, called trichroism (with only two observed in any one direction). In order to see pleochroism, the gem must be: colored (colorless gems transmit all colors of the spectrum of white light), a single crystal (an aggregate of crystals would scatter the light, obscuring the pleochroism), fairly transparent (numerous inclusions would again scatter the light), and viewed in some direction other than parallel to an optic axis. Pleochroism can be detected with the polariscope, although the calcite dichroscope is the preferred gem instrument to view pleochroism. The dichroscope is a metal tube with an opening at one end and lens at the

other. Optical calcite is mounted inside the tube, producing a double image of the square opening. When the gem is held over a bright light source and viewed through the dichroscope, each image has a different color indicating the vibration directions are at right angles and vary in wavelength. When two differing colors are detected, confirm pleochroism by rotating the instrument 90 degrees and the two colors should switch sides on the split image.
Trichroism can be detected by changing the orientation of the stone and one new color will be detected alongside one of the colors seen in the other orientation.

Refractometer for Testing Refraction

An instrument that measures the refractive index is a refractometer. The refractive index is actually read off a numeric scale, which shows a boundary line between a shaded and brighter portion. This line is a measure of the visible light rays striking the gem at the critical angle or where the light is totally reflected back into the opposite quadrant in the instrument, rather than reflected and refracted out of the stone.

The refractive indices of the refractometer's glass semicylinder (or the surface of the stone is placed on) and

the contact liquid (a liquid between the stone and the semicylinder to provide a contact without any outside interference) determine the upper limit of indices that can be read as

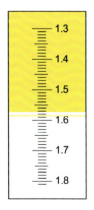

1.81. The most precise readings will be obtained using a monochromatic light source, or sodium light. One reading is possible in singly refractive stones, while two readings are possible when the stone is doubly refractive. The removable polarizing filter, when placed over the eyepiece of a refractometer, can show both refractive indices in doubly refractive stones, by rotating the filter (not the stone) if the orientation of the stone and transmission direction are aligned.

A reflectivity meter can be used with gemstones having of any refractive index, even over 1.81.

UV LW/SW Light for Testing Luminescence/Fluorescence

Gems that emit visible light after exposure to short-wave or long-wave ultraviolet radiation, they are said to be luminescent, or more specifically, fluorescent. A gem is phosphorescent if the luminescence continues after the UV light source has been removed. This  phenomenon of fluorescence and/or phosphorescence is the result of UV radiation absorbed by impurities or structural defects within the crystal structure, resulting in an oscillation of electrons between energy levels, and transmission of visible light.

Fluorescence produces vivid colors when an ultraviolet or invisible light source is directed at a gemstone. Where did this term come from? In the middle of the nineteenth century, Professor George Stokes was experimenting with sunlight and the mineral fluorite. He was in a darkened room and admitted sunlight through a small hole in a window shutter. When sunlight fell on a colorless fluorite specimen, he noticed the mineral displayed a bright blue color. Stokes coined the term fluorescence to describe the phenomenon he observed from the fluor-spar or fluorite, which he said was analogous to opalescence observed in opals. Therefore, fluorite is the basis for the word fluorescence but not all fluorite have fluoresces! The fluorescence can be a bold green, orangish-red, or whitish blue, and vary in intensity. Fluorescence may be unpredictable because some gems will have no reaction to the UV light source. Fluorescence testing involves cleaning the stone to be tested and then locating a very dark area for observations. Never look at the UV light source directly, as permanent damage to your eyes can occur. Place the gem on a surface when testing (not between metallic tweezers or your fingers). Stones that look purplish are inert, with the "purple" color being a reflection of the light source.

Color Filter for Testing Unique Luminescence

A color filter, also called an emerald filter or "Chelsea" filter, can help separate some natural, synthetic and imitation gem materials. Because color is due to absorption and transmission of different wavelengths of white light, the resulting green, for example, could be a mixture of different wavelengths. These different wavelengths can help distinguish a chromium colored emerald from other green stones and glass imitations, colored by other means. The emerald filter absorbs visible light, except for the long red wavelengths, which are transmitted,

causing the emerald to appear red under the filter. Unfortunately, for natural emerald the test is not conclusive, as some synthetics and natural emeralds will not react. Cobalt colored, synthetic spinel, a common aquamarine and topaz imitation, will also appear red with the filter, whereas the natural gems will appear greenish. An imitation of turquoise, dyed blue howlite, may react red, but natural and synthetic turquoise will appear greenish.

Part 8　Words and Expressions

apparatus[ˌæpəˈreɪtəs]	n. 器械,设备,仪器
brooch[bruːtʃ]	n. 胸针,领针
microscope[ˈmaɪkrəskəup]	n. 显微镜
triplet[ˈtrɪplɪt]	n. 三层石
phosphorescence[ˌfɔsfəˈresns]	n. 磷光
index[ˈɪndeks]	n. 指数
magnification[ˌmæɡnɪfɪˈkeɪʃən]	n. 放大倍率
lens[lenz]	n. 透镜,镜头
polariscope[pəuˈlærɪskəup]	n. 偏光镜
luminescence[ˌluːməˈnesns]	n. 发光
spectroscope[ˈspektrəskəup]	n. 分光镜
dichroscope[ˈdaɪkrəskəup]	n. 二色镜
pleochroism[plɪˈɔkrəuˌɪzəm]	n. 多色性
filter[ˈfɪltə]	n. 滤光片
mini[ˈmɪnɪ]	n. 迷你型
colourless[ˈkʌləlɪs]	adj. 无色的
refractory[rɪˈfræktərɪ]	adj. 难控制的,难熔的
senior[ˈsiːnjə]	adj. 高级的
priority[praɪˈɔrɪtɪ]	n. 优先权
interact[ˌɪntərˈækt]	vi. 互相作用
constant[ˈkɔnstənt]	n. 恒量
bench type polariscope	台式偏光镜
prism[ˈprɪzəm]	n. 棱镜,棱柱
refractometer[ˌriːfrækˈtɔmɪtə]	n. 折射仪
counterpart[ˈkauntəpɑːt]	n. 配对物
wavelength[ˈweɪvleŋθ]	n. 波长
bunch[bʌntʃ]	n. 串,束　v. 捆成一束
diffraction[dɪˈfrækʃn]	n. 衍射
grating[ˈɡreɪtɪŋ]	n. 光栅
instrument[ˈɪnstrumənt]	n. 仪器,器械
immersion[ɪˈməːʃn]	n. 沉浸
relief[rɪˈliːf]	n. 减轻

ethyl alcohol	n. 普通酒精
acetone[ˈæsɪtəun]	n. 丙酮
glycerine[ˈɡlɪsəri:n]	n. 甘油,丙三醇
olive[ˈɔlɪv]	n. 橄榄树,橄榄叶,橄榄枝,橄榄色
xylene[ˈzaɪli:n]	n. 二甲苯
clove oil	n. 丁香油
ethylene[ˈeθɪli:n]	n. 乙烯,乙烯基
bromide[ˈbrəumaɪd]	n. 溴化物
ethylene bromide	溴化乙烯,二溴乙烷
bromoform[ˈbrəuməˌfɔ:m]	n. 溴仿,三溴甲烷
methylene [ˈmeθɪli:n]	n. 亚甲基
iodide[ˈaɪədaɪd]	n. 碘化物
methylene iodide[meθɪli:n,aɪədaɪd]	n. 甲醛
sink[sɪŋk]	vi. 沉下,(使)下沉
vanish[ˈvænɪʃ]	vi. 消失,突然不见
aplanatic[ˌæpləˈnætɪk]	adj. 等光程的,(透镜)消球差的
flint[flɪnt]	n. 燧石
lead[li:d]	n. 铅(元素符号 Pb)
cement[sɪˈment]	n. 水泥
convex[ˈkɔnˈveks]	adj. 表面弯曲如球的外侧,凸起的
silica[ˈsɪlɪkə]	n. 硅石,无水硅酸,硅土
soda[ˈsəudə]	n. 苏打,碳酸水
lime[laɪm]	n. 石灰
oxide[ˈɔksaɪd]	n. 氧化物
illuminating[ɪˈlju:mɪˌneɪtɪŋ]	adj. 照明的
binocular[baɪˈnɔkjulə]	adj. 双目并用的
stereoscopic[ˈsterɪəˈskɔpɪk]	adj. 有立体感的
versatile[ˈvə:sətaɪl]	adj. 通用的
ocular[ˈɔkjulə]	n. 目镜,眼睛
multiply[ˈmʌltɪplaɪ]	v. 乘
reinvest[ˈri:ɪnˈvest]	vt. 再投资于,再授给
compound[ˈkɔmpaund]	n. 混合物
continuously[kənˈtɪnjuəslɪ]	adv. 不断地,连续地
transmit[trænzˈmɪt]	vt. 传输,转送,传达,传导,发射,遗传,传播
	vi. 发射信号,发报
rheostat[ˈri:əˌstæt]	n. 可变电阻器
iris[ˈaɪərɪs]	n. 虹膜
isotropic[aɪsəuˈtrɔpɪk]	adj. 各向同性的
anisotropic[əˌnaɪsəuˈtrɔpɪk]	adj. 各向异性的

plane polarize	平面偏振动光
velocity[vɪˈlɔsɪtɪ]	n. 速度,速率
tetragonal[teˈtrægənl]	adj. 四面体的
orthorhombic[ˌɔːθəˈrɔmbɪk]	adj. 斜方晶系的
hexagonal[hekˈsægənəl]	adj. 六方晶系的
monoclinic[ˌmɔnəˈklɪnɪk]	adj. 单斜(晶系)的
triclinic[traɪˈklɪnɪk]	adj. 三斜晶系的
crystallographic[ˌkrɪstələˈɡræfɪk]	adj. 结晶学的
vibrate[vaɪˈbreɪt]	v. (使)振动,(使)摇摆
mutually[ˈmjuːtʃuəli]	adv. 互相地,互助
polarizer[ˈpəuləraɪzə]	n. 偏光器,起偏镜
ray[reɪ]	n. 光线
trichroism[ˈtraɪkrəuɪzəm]	n. 三色性
aggregate[ˈæɡrɪɡeɪt]	n. 集合体 v. 聚集,集合
transparent[trænsˈpɛərənt]	adj. 透明的,显然的,明晰的
numerous[ˈnjuːmərəs]	adj. 众多的
optic[ˈɔptɪk]	adj. 光学上的
orientation[ˌɔːrɪənˈteɪʃn]	n. 方位,定位
scale[skeɪl]	n. 刻度
boundary[ˈbaundəri]	n. 边界
critical angle	临界角
quadrant[ˈkwɔdrənt]	n. 象限
semicylinder[ˈsemɪˈsɪlɪndə]	n. 半圆柱体,六方柱
monochromatic[ˌmɔnəukrəuˈmætɪk]	adj. 单色的
sodium[ˈsəudɪəm]	n. 钠(元素符号 Na)
align[əˈlaɪn]	vi. 排列
ultraviolet[ˈʌltrəˈvaɪələt]	adj. 紫外线的
defect[dɪˈfekt]	n. 缺点
oscillation[ˌɔsɪˈleɪʃn]	n. 摆动,振动
electron[ɪˈlektrɔn]	n. 电子
shutter[ˈʃʌtə]	n. 关闭者,百叶窗
stoke[stəuk]	v. (使)大吃
analogous[əˈnæləɡəs]	adj. 类似的,相似的,可比拟的
opalescence[ˌəupəˈlesns]	n. 乳白光,蛋白色光
orangish-red	橙红色
whitish[ˈwaɪtɪʃ]	adj. 发白的,带白色的
permanent[ˈpəːmənənt]	adj. 永久的,持久的
tweezers[ˈtwiːzəs]	n. 镊子
purplish[ˈpəːplɪʃ]	adj. 紫色调的

chromium [ˈkrəʊmɪəm]　　　　　　　　n. 铬（元素符号 Cr）
cobalt [kəʊˈbɔːlt]　　　　　　　　　　n. 钴（元素符号 Co）
howlite [ˈhaʊlaɪt]　　　　　　　　　　n. 羟硅硼钙石

Part 9　Check Your Understanding

Exercise 1　Answer the following questions

(1) What is gem's refractive index ?
(2) What are usages of the polariscopes when identifying the gemstones?
(3) What is the gem's pleochroism?
(4) What parameter is determined by the refractometer?
(5) Why can color filter separate some natural, synthetic, and imitation gem materials?

Exercise 2　Fill in the blanks

(1) Please illustrate several kinds of gems identification instrument: _____, _____, _____, _____.
(2) Gem spectroscope is classified into _____ and _____.
(3) Immersing the gem in a liquid with a different refractive index can make the gem _____.
(4) The function of a hand held lens or the microscope is to _____ _____ when identifying the gemstones.
(5) A gem is phosphorescent if the luminescence _____ after the UV light source has been removed.

Exercise 3　Translation

(1) Bench type polariscopes are used to quickly and easily check if gemstones are single or double refracting.
(2) A different method used to determine the refractive index is with immersion. Immersing gem material in a liquid with approximately the same refractive index, will make the gem nearly invisible. Immersing the gem in a liquid with a different refractive index can make the gem stand out clearly or in high relief.
(3) In order to see pleochroism, the gem must be: colored (colorless gems transmit all colors of the spectrum of white light), a single crystal (an aggregate of crystals would scatter the light, obscuring the pleochroism), fairly transparent (numerous inclusions would again scatter the light) and viewed in some direction other than parallel to an optic axis.
(4) 手持放大镜和宝石显微镜都可以对宝石进行放大观察。
(5) 红宝石是双折射宝石，而红色尖晶石是单折射宝石，用折射仪、偏光镜和二色镜都可以将它们鉴别开。
(6) 合成蓝色尖晶石在滤色镜下显红色调，而天然蓝色尖晶石在滤色镜下显绿色调。

Unit 5　Synthetic Gems Introduction

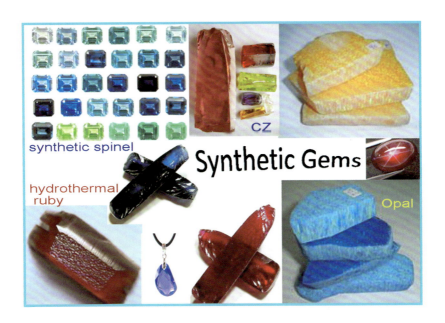

Part 1　Dialogue

Mary: Welcome to our shop. Can I help you?
Kate: I wish to buy a diamond ring.
Mary: How many carats would you like it to be?
Kate: I want one carat.
Mary: Is this one suitable for you?
Kate: No, it seems too old-fashioned to me.
Mary: How about this?
Kate: Let me try it on. Oh, it's too small for me. Haven't you got any larger ones?
Mary: Then you may take that one. It's very nice and the latest in style.
Kate: This fits me well. How much do you charge for it?
Mary: Eight thousand and two hundred dollars.
Kate: It's too expensive. I can only pay you seven thousand dollars.
Mary: I told you before, lady, our shop doesn't ask two prices.
Kate: Good, I'll have it. Have you got any brooches?
Mary: With diamond, ruby or sapphire?
Kate: Sapphire, please. How much is it?
Mary: Four hundred.

本章音频

Kate: All right, how much will it be altogether? Please send it to my address, I'll pay the messenger on delivery.
Mary: Sorry. It's our rule never to supply goods cash on delivery.
Kate: Well then, I'll give you a check for the amount on the Bank of China.

Part 2　Fill the Chinese Meanings with the Teacher's Tutorship

synthetic _____　　moissanite _____　　corundum _____　　anisotropic _____
flame-fusion _____　　cubic zirconium _____　　rutile _____　　inclusion _____
alumina _____　　strontium titanate _____　　imitation _____　　simulant _____

Part 3　Link the Relevant Words between Column A and Column B

column A	column B
moissanite	peridot
citrine	feldspar
emerald	SiC
cat's-eye stone	quartz
olivine	beryl
moonstone	chrysoberyl

Part 4　Useful Phrases

skills of pearl jewelry designing　珍珠首饰设计技巧
color dyeing of ancient jade　古玉的染色
China jewelry import and export　中国珠宝首饰进出口状况
Africa diamond mining area secrets probed　南非钻况探秘
flux-melt technique　助熔剂法
skull melting process　冷坩埚法
hydrothermal process　水热法
synthesis of gem-quality diamond　宝石级钻石的合成
synthetic cubic zirconia(CZ)　合成立方氧化锆
flame fusion technique　焰熔法
ruby boule　红宝石梨晶
a loose 0.95 carat diamond　一颗0.95克拉的裸钻
seed crystal　籽晶
Gemological Association of China（GAC）　中国宝玉石协会
National Gems and Jewelry Technology Administrative Center　自然资源部珠宝玉石首饰管理中心

Part 5 Some Sentences

(1) Some of the natural gems like ruby, sapphire, emerald and diamond are synthesized in the laboratories and manufactured by simulating the natural ones.

(2) Beautiful spinels were first produced in 1926, and now synthetic spinels with a variety of colors are on the market. The process is the same as in corundum. Since it is isotropic, it does not show dichroism, which distinguishes it from synthetic or natural corundum and natural zircon and chrysoberyl.

(3) This ruby by the split boule is top grade flawless synthetic corundum. There will be no cracks or blemishes in this material. We are selling this synthetic rough by the split boule instead of by gram. Average split boule size is 28 grams. This is simpler all around and worst case is you might get a 24 gram split boule. The price averages out to 40 cents per gram.

Part 6 Short Paragraph

Identification of Synthetic Gems

A synthetic gemstone is identical to a natural gemstone in almost every way. This includes the same basic crystal structure, refractive index, specific gravity, chemical composition, colors and other characteristics. Since the same gemological tests are used for stone identification on both natural and synthetic gems, it is sometimes even possible for a gemologist to be puzzled as to whether or not a stone is natural or synthetic. When this occurs, the best course of action is to send the stone to an accredited gem laboratory, like the Gemological Institute of America. They can positively determine whether a stone is synthetic or naturally occuring. Only minor internal characteristics allow separation of a synthetic gemstone from a natural gemstone.

It is very difficult to distinguish between natural and synthetic colorless sapphires. The natural crystals have microscopic irregularly-shaped gas and liquid inclusions, whereas the synthetic gems usually show microscopic cracks along and normal to the intersection of facets. Under the microscope, colored synthetic stones show curved lines parallel to the upper growth surface of the boule. They represent uneven distribution of pigmentation. Occasionally, especially in blue sapphire, they are visible to the unaided eye.

Synthetic Moissanite

Moissanite is composed of silicon carbide, a hard substance rated 9.25 on the Mohs scale. Small samples have been discovered in meteorites, but the moissanite that's sold for use in jewelry is lab grown.

Moissanite is a natural occurring mineral that is named after its discoverer, 1906 Nobel Prize (chemistry) winner Henri Moissan (1852—1907). As a natural source, this mineral is not suitable to be cut into gemstones as it occurs in too small grains and is very rare. Moissanite (silicon carbide) was synthesized (1893) before it was discovered (1905) in nature.

With refractive indices of 2.648 and 2.691, a dispersion of 0.104, a hardness of 9.25 on the Mohs scale and a specific gravity of 3.22, synthetic moissanite is much closer to diamond in overall appearance and harder than any previous diamond imitation. The thermal properties of synthetic moissanite are also so close to those of diamond that the thermal probes currently on the market react to synthetic moissanite as if it were "diamond". Synthetic moissanite can be easily separated from diamond by the doubling of facets, due to its high birefringence. Other clues are parallel needles (along the optic axis) and pinpoint inclusions in synthetic moissanite.

Part 7 Text

Synthetic Gems Introduction

Synthetic gems resemble very much the natural stones as they have the same chemical constitution. The synthetic ruby and natural ruby consist of Al_2O_3 and have the same physical properties as regards hardness, specific gravity, cleavage, refractive index and so on. But certain incidental and characteristic features are found helpful in the identification of synthetic stones and differentiating them from natural stones.

Even though there were a few early hard works, successful gem synthesis began in the late 1800s with the development of synthetic ruby by the flame-fusion method. The need for a variety of high-quality crystals for a range of industrial applications (i.e., optics components, laser

Unit 5 Synthetic Gems Introduction

crystals and other uses) and the increased awareness of crystal growth mechanisms, have since combined to yield the production of a wide variety of crystals, some of which are synthetic counterparts of natural gemstones. A synthetic gem material has the similar chemical composition and crystal structure as a natural gem mineral. In contrast, a gem imitation has the appearance of a natural gemstone, but has a different chemical composition, physical properties and crystal structure. Both synthetic gem materials almost always possess gemological properties that allow them to be distinguished from the corresponding natural gemstones.

Current methods used for gem synthesis fall into two general categories: crystallization from a fluid of different composition (such as a flux or aqueous hydrothermal solution) and crystallization from a melt with roughly the same chemical composition as the crystal being grown. Synthetic diamonds, grown from a metallic flux at high temperatures and pressures, have evoked concern in the jewelry trade ever since the first production of gem quality crystals in the early 1970s. However, due to the growth conditions involved, as well as the great expense and restricted availability of the growth equipment used, the actual number of gem-quality synthetic diamonds continues to be limited. Those that are encountered in the jewelry trade are mainly brownish-yellow crystals, weighing 1 carat or less (1 carat=0.2 g), which can yield faceted pieces of 0.5 carat or smaller in size. In contrast to

synthetic diamonds, colorless or near colorless imitation materials are much more abundant.

Over the years, a number of natural gem minerals and synthetic materials have been used to imitate colorless diamond. Cubic zirconium oxide (cubic zirconium or CZ) is the most widespread similitude because of its low cost and similar appearance to a polished diamond. It can be readily distinguished from diamond on the basis of a difference in heat conduction, which can be detected with a simple gemological test instrument. Within the past two years, a new material, synthetic moissanite (silicon carbide), has been marketed for jewelry purposes. It has caused some serious identification problems among jewelry, because it cannot be distinguished from diamond by the heat conduction test mentioned earlier. However, synthetic moissanite displays optical features due to its anisotropic optical character (seen as an optical doubling when viewed with magnification) which allow it to be easily recognized by trained gemologists.

Among colored stones, the most important synthetics are synthetic corundum (sapphire and ruby), emerald, spinel and amethyst. As indicated they are produced by both solution and melt crystallization techniques. In the marketplace, flame-fusion and crystal-pulled synthetics are relatively less expensive and therefore are more abundant than the flux and hydrothermal synthetics. Solution-grown synthetic amethyst falls into the same less-expensive category because of the existing large-scale production facilities for synthetic quartz for using in the electronics industry. In contrast, hydrothermal and flux-grown ruby, sapphire and emerald are considered as "luxury" synthetics that command higher prices.

Method for producing synthetic rubies and sapphires: Originally developed (1902) by a French chemist, Auguste Verneuil, the process produces a boule (a mass of alumina with the same physical and chemical characteristics as corundum) from finely ground alumina (Al_2O_3) by means of an inverted oxyhydrogen torch that opens into a ceramic muffle. With slight modifications, this method is used to produce spinel, rutile and strontium titanate.

Highly purified alumina is placed in a container with a fine sieve at its base. When the container is tapped by a mechanically-activated hammer, the alumina sifts down into the enclosed chamber. Oxygen passes into this chamber and carries the fine alumina particles into the intense heat of the central part of an oxyhydrogen flame, where they fuse and fall on the molten upper surface of the boule as droplets. Flame characteristics and the rate of powder feed and boule lowering are adjusted to produce a boule of uniform diameter. The temperature of the upper surface of the boule is held just above the melting point, which for colorless sapphire is 2030℃ (about 3690°F). When a boule reaches the desired size, normally 150 to 200 carats, the furnace is shut down and the boule is cooled.

Strain develops during cooling, because the outer surface cools faster than the interior; this phenomenon causes considerable loss from cracking during the manufacturing process. The strain is relieved by splitting the boule longitudinally. Some residual strain is not disadvantageous for gem and most industrial uses is left in the half-boule developed by splitting. Strain-free, whole boule may be produced by annealing at 1950℃.

Part 8 Words and Expressions

fashioned[ˈfæʃənd]	adj.……式的,……风的
delivery[dɪˈlɪvərɪ]	n. 递送,交付,分娩,交货,引渡
flame-fusion[fleɪm][ˈfjuːʒən]	焰熔法
cubic zirconium oxide[ˈkjuːbɪk][zɜːˈkəʊnɪəm][ˈɒksaɪd] 立方氧化锆	
alumina[əljuːmɪnə]	n. 氧化铝(亦称矾土)
simulant[ˈsɪmjulənt]	n. 仿制品
olivine[ˌɒlɪˈviːn]	n. 橄榄石
moonstone[ˈmuːnstəʊn]	n. 月光石
chrysoberyl[ˈkrɪsəberɪl]	n. 金绿宝石
bubble[ˈbʌbl]	n. 气泡
loose[luːs]	n. 放任,放纵 adj. 宽松的,不精确的,不牢固的,散漫的,自由的 vt. 释放,放枪,开船
	vi. 变松,开火 adv. 松散地
simulate[ˈsɪmjuleɪt]	vt. 模拟,模仿,假装,冒充
isotropic[ˌaɪsəˈtrɒpɪk]	adj. 等轴的
boule[buːl]	n. 梨形人造宝石
blemish[ˈblemɪʃ]	n. 瑕疵,污点,缺点
identical[aɪˈdentɪkəl]	adj. 同样的
accredited[əˈkredɪtɪd]	adj. 公认的,质量合格的
positively[ˈpɒzətɪvlɪ]	adv. 肯定地
minor[ˈmaɪnə]	adj. 较小的,次要的
represent[ˌreprɪˈzent]	vt. 象征
uneven[ʌnˈiːvən]	adj. 不平坦的
distribution[ˌdɪstrɪˈbjuːʃn]	n. 分配,分发
pigmentation[ˌpɪgmənˈteɪʃn]	n. 染色,着色
unaided eye	肉眼
silicon[ˈsɪlɪkən]	n. 硅(元素符号 Si)
carbide[ˈkɑːbaɪd]	n. 碳化物
meteorite[ˈmiːtɪəraɪt]	n. 陨石
overall[ˈəʊvərɔːl]	adj. 全部的,全面的
birefringence[ˌbaɪrɪˈfrɪndʒɪns]	n. 双折射
needless[ˈniːdlɪs]	adj. 不需要的,不必要的
optic axis[ˈɒptɪk][ˈæksɪs]	光轴
resemble[rɪˈzembl]	vt. 像,类似

incidental[ˌɪnsɪˈdentl]	adj. 附带的,伴随的,非主要的,偶然的,容易发生的
	n. 伴随事件
mechanism[ˈmekənɪzəm]	n. 机制
corresponding[ˌkɔrɪsˈpɔndɪŋ]	adj. 相应的
fluid[ˈfluːɪd]	n. 流体,液体
flux[flʌks]	n. 熔剂,助熔剂
aqueous[ˈeɪkwɪəs]	adj. 水成的
hydrothermal[ˈhaɪdrəuˈθəːməl]	adj. 热液的
solution[səˈluːʃn]	n. 解决方案
evoke[ɪˈvəuk]	vt. 唤起
conduction[kənˈdʌkʃn]	n. 传导
heat conduction	热导率
synthetics[sɪnˈθetɪks]	n. 人工合成材料
crystal-pulling synthetics	提拉法合成晶体
luxury[ˈlʌkʃərɪ]	n. 奢侈,华贵
inverted[ɪnˈvəːtɪd]	adj. 反向的,倒转的
oxyhydrogen[ˌɔksɪˈhaɪdrədʒən]	n. 氢氧混合气　adj. 氢氧混合的
torch[tɔːtʃ]	n. 火把,启发之物
ceramic[sɪˈræmɪk]	adj. 陶器的
muffle[ˈmʌfl]	v. 包,蒙住,压抑(声音)　n. 围巾,头巾,消声器
sieve[sɪv]	n. 筛,滤网
mechanically-activated hammer[mɪˈkænɪkəlɪ][ˈæktɪveɪtɪd][ˈhæmə]	n. 频锤
chamber[ˈtʃeɪmbə]	n. 容器,室
furnace[ˈfəːnɪs]	n. 炉子,熔炉
strain[streɪn]	n. 张力,应变,过度的疲劳,紧张
crack[kræk]	n. 裂缝,噼啪声
longitudinal[lɔndʒɪˈtjuːdɪnl]	adj. 经度的,纵向的
elongate[ˈiːlɔŋgeɪt]	v. 拉长
stem[stem]	n. 茎,干
residual[rɪˈzɪdjuəl]	adj. 剩余的,残留的

Part 9　Check Your Understanding

Exercise 1　Answer the following questions

(1) What is synthetic gemstone?

(2) What is the price for synthetic ruby boule?

(3) Please describe the gemological characters of the synthetic moissanite.

(4) What is the way of identifying the synthetic gem from its counterparts of natural gemstones?

(5) Please describe the general situation of synthetic rubies and sapphires by flame-fusion

Unit 5 Synthetic Gems Introduction

method.

Exercise 2 Fill in the blanks

(1) Please illustrate several synthetic gems: _____, _____, _____, _____, _____.

(2) GAC is the breviary for _____.

(3) GIA is the breviary for _____.

(4) The successful gem synthesis began in the late 1800s with the development of synthetic ruby by _____ method.

(5) Synthesizing rubies by flame-fusion method, strain develops during cooling, because _____.

Exercise 3 Translation

(1) We are selling this synthetic rough by the split boule instead of by gram. Average split boule size is 28 grams. This is simpler all around and worst case is you might get a 24 gram split boule. The price averages out to 40 cents per gram.

(2) A synthetic gemstone is identical to a natural gemstone in almost every way. This includes the same basic crystal structure, refractive index, specific gravity, chemical compositions, colors and other characteristics.

(3) Among colored stones, the most important synthetics are synthetic corundum (sapphire and ruby), emerald, spinel and amethyst. As indicated they are produced by both solution and melt crystallization techniques.

(4) 合成莫依桑石比钻石的其他仿制品具有更高的硬度。

(5) 鉴别红宝石与合成红宝石有难度。

(6) 宝石学家可以通过放大观察鉴别出合成莫依桑石,因为它是光学非均质体。

Unit 6　Gemstone Treatments and Enhancements

Part 1　Dialogue

Mary：Good afternoon, something for you?
Kate：Yes, thanks. I bought this jadeite brooch last week, but I want to exchange it for a bracelet.
Mary：Why?
Kate：My husband said that this brooch does not fit me.
Mary：Oh, have you carried the receipt?
Kate：Yes, here you are.
Mary：The bracelets is possibly more expensive than the brooch.
Kate：It doesn't matter, I can compensate the price difference with cash.
Mary：Which bracelet do you like?
Kate：The one with light purple.
Mary：Sorry, someone has already spoken for it.
Kate：What a pity! What about that one on the side?
Mary：Oh, the price difference is 800 RMB.
Kate：Yes, I'll take it.
Mary：Please foot this bill on the counter, I'll pack the bracelet

本章音频

for you.
Kate: Thank you, see you later.
Mary: Bye-bye.

Part 2 Fill the Chinese Meanings with the Teacher's Tutorship

heating _____   bleaching _____   waxing _____   filling _____
dyeing _____   irradiation _____   laser drilling _____   coating _____
diffusion _____   filling _____   treatment _____   enhancement _____

Part 3 Link the Relevant Words between Column A and Column B

column A	column B
dyeing jadeite	likely irradiation
black pearls	enhancement
blue topaz	treatment
irradiated tourmaline	chemicals high-temperature heat treatment
heated rubies	tahitian
diffusion	C-jade

Part 4 Useful Phrases

appearance of a gem   宝石的外观
heart-shaped pendant   心形吊坠
Shanghai International Jewelry Fair   国际珠宝展览会
small retail jewelry shop   小型珠宝零售商店
a piece of jewelry   一件首饰
carving wax   雕蜡
fine gold (FG)   纯金
Fellowship of Gemological Association and Gem Testing Laboratory of Great Britain (FGA)
        英国宝石协会和宝石检测实验室
Diamond High Council (HRD)   比利时钻石高阶层议会
Central Selling Organization (CSO)   中央统售机构
International Gemological Institute (IGI)   国际宝石学院
International Colored Gemstone Association (ICA)   国际有色宝石协会
claw setting amethyst   爪镶的紫水晶
freshwater cultured pearl   淡水养殖珍珠
dyeing jadeite   染色的翡翠

Part 5 Some Sentences

(1) The term "enhancement" is defined to be any treatment process other than cutting and polishing that improves the appearance(color/clarity/phenomena), durability, or availability of a gemstone.

(2) Many people confuse gemstone enhancements with the terms synthetic, man-made, lab-grown or other processes which involve the creation of materials not found in the earth.

(3) The fact is, most gemstones used in jewelry have been treated to improve their appearance. Treated gemstones can be a good choice, because they should be more affordable than untreated stones that achieve their quality naturally.

Part 6 Short Paragraph

Symbols for Specific Forms of Enhancement

B = Bleaching: The use of heat, light and/or chemicals or other agentia to lighten or remove a gemstone's color. This is often accompanied by subsequent dying and/or impregnation. Example: bleached cultured pearl; bleached/impregnated jadeite("B-jade").

C = Coating: The use of such surface enhancements as lacquering, enameling, inking, foiling, or sputtering of films to improve appearance, provide color or add other special effects. Example: coated diamond.

D = Dyeing (staining): The introduction of coloring matter into a gemstone to give it new color, intensify existing color or improve color uniformity. Example: dyed green jadeite.

F = Filling: The filling of surface-breaking cavities or fissures with colorless glass, plastic, or some similar substance. This process will improve durability, appearance and/or add weight. Example: ruby.

Fh = Flux healing: During heat enhancement, fluxes (or heat alone) may be used to heal fractures/fissures which were formerly open. The process dissolves the walls of the fractures

and redeposits the molten gem material, healing the fractures closed. Example: ruby (particularly that from Möng Hsu, Burma).

H = Heating: The use of heat to alter color, clarity, and/or phenomena. Example: ruby, sapphire, tanzanite, aquamarine, demantoid garnet.

I = Impregnation: The impregnation of a porous gemstone with a colorless agentia (usually plastic) to give it durability and improve appearance. Example: stabilized turquoise.

L = Lasering: The use of a laser and chemicals to reach and alter inclusions. Example: diamond.

O = Oiling/resin infusion: The filling of surface-breaking fissures with a colorless oil, wax, resin or other colorless substances, except glass or plastic, to improve the gemstone's appearance. Example: emerald.

R = Irradiation: The use of neutrons, gamma ray, ultraviolet and/or electron bombardment to alter a gemstone's color. The irradiation may be followed by a heating process. Example: blue topaz.

U = Lattice ("bulk" or "surface") diffusion: Outside-in diffusion of coloring chemicals via high-temperature heat treatment to produce color and/or asterism. Example: lattice diffusion-treated sapphire.

W = Waxing/Oiling: The impregnation of a colorless wax, paraffin and/or oil in porous gemstones to improve appearance. Example: jadeite.

Part 7 Text

Gemstone Treatments and Enhancements

The treatment and enhancement of gemstones has existed for hundreds and hundreds of years. The first documentation of treatments was presented by Pliny. And, 2000 years later, many of these treatments are still being used today! Some enhancements improve on nature, cannot be detected and are permanent; this provides the gem market with a larger supply of beautiful gemstones. Other treatments produce dramatic changes in the gemstone itself or it's clarity; the irradiation and heating of colorless topaz that permanently transforms it into blue topaz is an excellent example. A few treatments are less stable and should be avoided by the knowledgeable buyer. Following is a description of some common treatments. This is just the tip of the iceberg.

In the past, treatments of gemstones were usually done by the cutter. The lapidary wanted the value of the finished product to be as high as possible. Today, there are centers, such as Bangkok in Thailand where there are facilities that specialize in treatment of both rough and fashioned gems. The heat treatment of corundum (rubies and sapphires) is an excellent example. The heat treatments of corundum (both simple heating and heating with a flux, such as beryllium) are often done before cutting, and may not be disclosed to the lapidary before cutting is done.

Treatments and Pricing

There are some gemstones that would not even exist if it were not for treatments. The abundance of citrine, in shades of yellow, gold and orange is the result of heat treating amethyst. Naturally occurring citrine is quite rare in nature. If it was not for treatments, the stone would be far more expensive than it is!

Tanzanite in shades of violet and blue depends on heat treatment to produce enough supply to meet the demands of the public.

Pink topaz is another example of a gem that would not be available without heat treatment. Not only are these treatments acceptable, they are necessary to keep these products affordable and available.

Recent demand for unheated sapphires and rubies has caused a price increase of as much as 50%~100% for unheated material. Does this mean that the untreated gem is more beautiful? NO! In most cases the heating enhances the gemstone to make it more beautiful; the price premium is the result of the rarity of being unheated!

Heating

Heating is the most common treatment available. It can cause the color of a stone to lighten, darken, or change completely. It can bring about an improvement in clarity and brightness. Heating is detectable only by trained observers in a laboratory setting and is usually irreversible under normal conditions. Unheated rubies and sapphires will contain microscopic rutile needles or tiny gas bubbles in pockets of liquid which are evidence that laboratories can use to guarantee that these stones have not been heated. If these gems are the finest color, they will command premium prices due to their extreme rarity.

The following gems are routinely heat treated: tanzanite, citrine, pink topaz, aquamarine, Paraiba tourmaline, apatite, ruby, sapphire, zircon (both blue and colorless).

Oiling

Oiling of emerald is universal, but not every emerald is oiled. (fine untouched specimens will command astronomical prices). When the rough emerald is mined, it is thrown into a barrel of oil; when it is cut, oil is used as a lubricant on the cutter's lap. The colorless oil seeps into the fissures on the surface of the emeralds. When the fractures contain the oil, they are less eye visible. To complete this process, oil is pressurized into the fissures of the polished stone. This is something that must be accepted; it's the way it is! The only way you will find an emerald that isn't oiled is if there are no fractures at the surface of the emerald, so no oil can get inside the stone. If color is equal, obviously you will pay more for an emerald if it has no fissures that reach the surface; they simply will have fewer inclusions. If an emerald that originally had fissures that reached the surface, is put into an ultrasonic or is steamed clean, then the oil may be leached out and fractures. This will make the surfacing inclusions appear whiter and more obvious. In this case, the stone can be re-oiled.

Recently, I have read articles that other colored stones such as rubies, alexandrite, other varieties of chrysoberyl, and demantoid garnets have been treated with oils and resins to make surfacing inclusions less visible. Occasionally colored oils are used on emeralds and rubies. The idea is to add color while concealing fractures. You want to avoid buying these because you can't judge the true color or know how bad the fractures are. This is done to deceive the buyer. Fortunately this is not common and it is unlikely you will encounter this if you buy from a reputable source in the United States. Synthetic resins can be used to fill in fractures in emeralds and other stones with fractures that reach the surface of the gem. Hardeners are often applied to make the process more permanent. The use of these resins, with hardeners is NOT acceptable treatment.

Irradiation

Irradiation means pounding material with subatomic particles or radiation. Sometimes irradiation is followed by heating to produce a better or new color for the gem. Blue topaz is the most common example. Although blue topaz occurs in nature, it is quite rare and pale in color. In the United States irradiated gems are regulated by the Nuclear Regulatory Agency to in an attempt to insure there is no harmful residual radiation.

You do not have this protection if you buy it out of this country. Today irradiation of blue topaz has created shades not found in natural blue topaz; prices are very reasonable for irradiated blue topaz since there is a great deal of competition in the wholesale end of this market. If you could find an untreated blue topaz, it would sell for a price comparable to untreated Imperial Topaz. Tourmaline can be irradiated to darken pink stones into red ones; these are indistinguishable from natural red ones. Off colored diamonds can be irradiated and heated and turned into intense greens, yellows, blues, browns & pinks. These stones are fairly common. Irradiated diamonds will sell for much less per carat than the naturally colored ones of comparable color, clarity grade, and size. Cultured pearls can be irradiated to produce gray or blue colors; but dyeing in these colors is more common. Irradiated pearls will sell for about the same price as the dyed pearls, this should be well below the prices asked for pearls with very fine colors. Varieties of quartz and spodumene are irradiated and subsequently annealed with heat to produce dramatic and desirable colors.

Dyeing

Without dyeing there would be no black onyx; this is not a natural color of chalcedony!

Chalcedony or more commonly known as agate, is often dyed blue, green, or orange and carved into bowls, statues, or cut into beads. This is fine, as there are some lovely pieces around using this stuff, especially carved animals and the like and no one minds that it's not "natural". Japanese cultured pearls, which are grown in an Akoya oyster that produces pearls up to about 10 millimeters, grow into a limited selection of colors with various overtones of colors. If they are dark gray, bluish, violet, nearly black, or intense bronze, assume they are dyed. To meet current demand for pearls with rose overtones, some cultured pearls have been given a pink tint; this can be detected by looking for concentrations of dye around drill holes or around blemishes. On the other hand, South Sea cultured pearls which are generally larger than the Japanese cultured pearls, may grow into a variety of exotic colors naturally because they are grown in a different variety of oyster.

Bleaching and Coating

Bleaching is a process for organic gem materials such as ivory, coral, and for pearls and cultured pearls. It lightens the color and is permanent and undetectable. No price difference exists as a result.

Coating is a process used and described for over 200 years! where a lacquer or film of some type is applied to improve a gem's appearance. Today, coatings are increasingly utilized to alter and improve the color of gems. Mystic topaz is an example of a coated gem that was conceived by Azotic Coating Technologies. The company is now coating topaz in all colors, including pinks and rich "imperial" tones. Recent reports have indicated that tanzanite is showing up in the labs with coatings on the pavilions to improve the appearance of saturation. Coatings are occasionally identified on diamonds to improve the apparent color of an off-colored stone and deceive a buyer.

Opals may have a black coating on the back to intensify the play of color or to give the appearance of a black opal; this can take the form of a simple coat of black lacquer or what is called a "doublet". A doublet is a thin layer of opal cemented to a black onyx base.

Diffusion

Diffusion was originally used on sapphires. Chemicals, like beryllium, were infused at high temperatures, and actually penetrated the gems. Early diffusion only produced color on the surface of the gem's surface and was referred to as "Surface Diffusion". Surface diffusion was easily detectable with immersion and often with simple magnification. Great advancements have been made in diffusion treatment in the last decade and it was discovered that if corundum is heated to very high temperatures for a long duration, the diffusion would penetrate the entire stone!

It can improve color, change color, or create asterism (stars).

Filling

Filling is used on gems with surface fractures or cavities. Glass, plastic or other materials are used to fill these holes. This is sometimes done to rubies. With close examination with magnification you may be able to spot differences in surface luster, or see a spectral effect in fractures when viewed with dark-field illumination. The AIGS, The Asian Institute of Gemological Sciences, has done extensive research on filled rubies.

Infilling Diamonds

Diamonds with inclusions are sometimes filled with glass to make them appear clearer. Oved and Yehuda Diamonds have undergone this treatment. Filler can be damaged by heat, ultrasonic cleaning and by re-tipping. The filling does not repair the inclusion, it just makes it less visible.

If you look at a filled diamond closely, rotate it under light, you should be able to notice a bluish flash. Both Yehuda and Oved will usually refill your diamond for free if it is ever damaged. Check for guarantees before buying such a diamond.

Lasering

Lasering is sometimes used on diamonds. The process drills very tiny holes into a diamond to provide access to an inclusion which detracts from the beauty of the stone. The inclusion can then be, vaporized or bleached to make it less obvious if it is not burned out by the lasering. Under magnification laser holes are visible when viewed at the correct angle. A lasered diamond would be classified in the slightly imperfect or imperfect category regardless of the improvement in apparent clarity and should be priced accordingly.

Gems That Are Not Enhanced

There are some gemstones that are not known to be enhanced. These include: garnets (with the exception of demantoid), peridot, iolite, spinel, varieties of chrysoberyl, tourmaline (with the exception of the Paraiba variety), malachite, hematite, and feldspar with the probable exception of varieties of andesine and labradorite. Keep in mind that new technology in gemstone treatment is always changing and improving and many are seriously difficult, if not impossible, to detect.

Part 8　Words and Expressions

bleach[bliːtʃ]	v. 漂白
wax[wæks]	n. 蜡
fill[fɪl]	vt. 填充
dye[daɪ]	n. 染料　vt. 染
irradiation[ɪˌreɪdiˈeɪʃn]	n. 放射,照射
laser[ˈleɪzə]	n. 激光
coating[ˈkəʊtɪŋ]	n. 涂层,镀层
diffusion[dɪˈfjuːʒn]	n. 扩散,漫射
treatment[ˈtriːtmənt]	n. 处理
enhancement[ɪnˈhɑːnsmənt]	n. 优化
jade[dʒeɪd]	n. 玉石

Tahitian[tɑːˈhiːtɪən]	adj. 塔希提岛的　n. 塔希提人
improve[ɪmˈpruːv]	v. 改善,改进
availability[əˌveɪləˈbɪlɪtɪ]	n. 可用性,有效性,实用性
confuse[kənˈfjuːz]	vt. 搞乱,使糊涂
involve[ɪnˈvɔlv]	vt. 包括
accompany[əˈkʌmpənɪ]	vt. 陪伴,伴奏
subsequent[ˈsʌbsɪkwənt]	adj. 后来的,并发的
impregnation[ˌɪmpreɡˈneɪʃn]	n. 注入
lacquering[ˈlækərɪŋ]	n. 上漆
enameling[ɪˈnæməlɪŋ]	n. 上釉,上涂料,上釉药
foil[fɔɪl]	n. 箔　vt. 贴箔于
sputtering[ˈspʌtərɪŋ]	n. 反应溅射法,飞溅,溅射,阴极真空喷镀,阴极溅镀,喷射,溅蚀
film[fɪlm]	n. 薄膜,膜层
stain[steɪn]	n. 污点,瑕疵
staining[ˈsteɪnɪŋ]	n. 着色
uniformity[ˌjuːnɪˈfɔːmɪtɪ]	n. 一致
cavity[ˈkævɪtɪ]	n. 洞,空穴
fissure[ˈfɪʃə]	n. 裂缝
deposit[dɪˈpɔzɪt]	n. 堆积物,沉淀物,存款,押金,保证金,存放物　vt. 存放,使沉积　vi. 沉淀
tanzanite[ˈtænzənaɪt]	n. 坦桑石
demantoid[dɪˈmæntɔɪd]	n. 翠榴石
stabilize[ˈsteɪbɪlaɪz]	v. 稳定
resin[ˈrezɪn]	n. 树脂
infusion[ɪnˈfjuːʒən]	n. 灌输
neutron[ˈnjuːtrɔn]	n. 中子
gamma ray[ˈɡæmə][reɪ]	n. 伽马射线,γ射线
ultraviolet[ˌʌltrəˈvaɪəlɪt]	adj. 紫外线的
electron[ɪˈlektrɔn]	n. 电子
bombardment[bɔmˈbɑːdmənt]	n. 炮击,轰击
asterism[ˈæstərɪzəm]	n. 星光效应
lattice[ˈlætɪs]	n. 晶格
paraffin[ˈpærəfɪn]	n. 石蜡
documentation[ˌdɔkjumenˈteɪʃn]	n. 文件
dramatic[drəˈmætɪk]	adj. 戏剧性的,生动的
clarity[ˈklærɪtɪ]	n. 净度
permanently[ˈpəːmənətlɪ]	adv. 永存地,不变地
transform[trænsˈfɔːm]	vt. 转换,改变,改造,使……变形

Unit 6　Gemstone Treatments and Enhancements

stable['steɪbl]	adj. 稳定的
avoid[ə'vɔɪd]	vt. 避免,消除
iceberg['aɪsbɜːg]	n. 冰山,冷冰冰的人
cutter['kʌtə]	n. 刀具,切割机
facility[fə'sɪlɪtɪ]	n. 设备,工具
specialize['speʃəlaɪz]	vi. 专攻,专门研究,使适应特殊目的,使专用于
beryllium[bə'rɪlɪəm]	n. 铍(元素符号 Be)
affordable[ə'fɔːdəbl]	adj. 可提供的,可给予的,可供应得起的
available[ə'veɪləbl]	adj. 可用到的,可利用的
brightness['braɪtnɪs]	n. 亮度,光亮,明亮,聪明,智慧
detectable[dɪ'tektəbl]	adj. 可发觉的,可看穿的
irreversible[ɪrɪ'vɜːsəbl]	adj. 不可逆的,不可改变的,不能撤回的,不能取消的
guarantee[gærən'tiː]	n. 保证,保证书,担保
premium['priːmɪəm]	n. 额外费用,奖金,奖赏,保险费
untouched['ʌn'tʌtʃt]	adj. 未触及的
astronomical[æstrə'nɔmɪkəl]	adj. 天文学的,庞大无法估计的
barrel['bærəl]	n. 桶
lubricant['luːbrɪkənt]	n. 滑润剂
seep[siːp]	v. 渗出,渗漏
pressurize['preʃəraɪz]	vt. 增压,密封
subatomic['sʌbə'tɔmɪk]	adj. 小于原子的,亚原子的,次原子的
particle['pɑːtɪkl]	n. 粒子,点,极小量,微粒,质点
wholesale['həulseɪl]	n. 批发　adj. 批发的,大规模的
indistinguishable[ɪndɪs'tɪŋgwɪʃəbl]	adj. 不能辨别的,不能区别的
cultured pearls	养殖珍珠
onyx['ɔnɪks]	n. 缟玛瑙
chalcedony[kæl'sedənɪ]	n. 玉髓
carve[kɑːv]	v. 雕刻,切开
statue['stætjuː]	vt. 以雕像装饰　n. 雕像
bead[biːd]	n. 珠子,水珠
overtone['əuvətəun]	n. 泛音,暗示,折光的色彩
concentrations[kɔnsn'treɪʃn]	adj. 集中的
ivory['aɪvərɪ]	n. 象牙
lighten['laɪtn]	v. 减轻
utilize['juːtɪlaɪz]	vt. 利用
mystic['mɪstɪk]	adj. 神秘的,神秘主义的
azotic[ə'zɔtɪk]	adj. 氮的
imperial[ɪm'pɪərɪəl]	adj. 皇帝的
pavilion[pə'vɪljən]	n. 亭部

penetrate[ˈpenɪtreɪt]　　　　　　　vt. 穿透,渗透
duration[djuˈreɪʃn]　　　　　　　　n. 持续时间,为期
entire[ɪnˈtaɪə]　　　　　　　　　　adj. 全部的,完整的,整个
illumination[ɪˌluːmɪˈneɪʃn]　　　　n. 照明,阐明,启发,灯彩(通常用复数)
ultrasonic[ˌʌltrəˈsɒnɪk]　　　　　adj. 超音速的,超声的
access[ˈækses]　　　　　　　　　n. 通路,访问,入门　vt. 存取,接近
detract[dɪˈtrækt]　　　　　　　　v. 转移
vaporize[ˈveɪpəraɪz]　　　　　　　v. (使)蒸发
regardless[rɪˈɡɑːdlɪs]　　　　　　adj. 不管,不顾
malachite[ˈmæləkaɪt]　　　　　　n. 孔雀石
hematite[ˈhemətaɪt]　　　　　　　n. 赤铁矿
andesine[ˈændɪziːn]　　　　　　　n. 中长石
labradorite[ˈlæbrədɔːraɪt]　　　　n. 闪光拉长石,富拉玄武岩

Part 9　Check Your Understanding

Exercise 1　Answer the following questions

(1) What is the purpose for treatment and enhancement of gemstones?

(2) How do you think for the price for treated sapphire?

(3) Which gemstones are routinely heated treatment?

(4) What is diffusion treatment of gemstones?

(5) When is the diamond filled with glass?

Exercise 2　Fill in the blanks

(1) Please illustrate several ways of treatment or enhancement: _____, _____, _____, _____.

(2) FGA is the breviary for _____.

(3) CSO is breviary for _____.

(4) IGI is breviary for _____.

(5) Heating can cause the color of a stone to _____, _____, or _____. It can bring about an improvement in clarity and brightness.

(6) Sometimes irradiation is followed by _____ to produce a better or new color for the gem.

(7) Opals may have a black coating on the back to intensify the _____ or to give the appearance of a black opal.

Exercise 3　Translation

(1) Heat and radiation is often used to change or enhance gemstone colors.

(2) Diffusion is used to deepen a gem's color. Diffusion only intensifies a gemstone's outer layers.

(3) Oil and waxes enhance gemstone colors by filling-in fine surface cracks, blending them away temporarily.

(4) Fracture filling coats gems with a clear or colored epoxy resin or another substance. The

treatment fills-in cracks, which improves the appearance of the gemstone.
(5) Laser drilling removes inclusions, improving clarity.
(6) 祖母绿常作浸油处理,这种方法可被人们接受,是优化。
(7) 扩散处理可以改善颜色、改变颜色甚至可以产生星光。
(8) 激光处理常用来改善钻石的净度。

Part 10 Self-Study Material

Dyeing Gems

Tahitian black pearls are a good example of naturally colored black pearls. Cultured pearls with a natural exotic color will command a much higher price than a dyed one. Dyeing of chalcedony and of pearls is prevalent, permanent, and acceptable. These colors do not occur in nature; no deception is involved. Dyeing of other materials, jade, lapis lazuli, turquoise, coral, ruby, emerald and sapphire may be less acceptable. Generally, dyeing of these materials is done to disguise poor quality goods. Dyed lapis lazuli can be easily tested by rubbing it with a piece of cotton soaked with acetone (fingernail polish remover). If it is dyed, blue color will eventually rub off on the cotton. Dyed lapis should be much less expensive than fine natural lapis. In the case of lapis lazuli or turquoise, the natural material is not that expensive, so why bother with inferior material unless it is irresistibly cheap or you just love the color? Dyed lapis lazuli may bleed blue onto the wearer or his or her clothing (not a fun thing to remove, trust me). Dyed jade may be tricky to detect, so be careful if the price seems "too good". An inexpensive tool (around $30) called a Chelsea Filter and supplement emerald filters can somewhat useful detecting dyed jade but the sophistication of the bleaching and polymer impregnation of jadeite can be extremely hard to detect without the aid of spectrographic analysis. Coral beads may also be dyed. Suspect coral that has a very intense color, coupled with an inexpensive selling price. I recently encountered strands of sapphire beads which were quench cracked and died. The treatment was easily visible with microscopic observation, but it did not bleed at all when soaked in acetone.

Unit 7 Gemstone Inclusion

Part 1　Dialogue

本章音频

Mary：Can I be of any assistance?

Kate：Mind yourself, I'll just look around.

Mary：Would you mind me recommending something?

Kate：Ok, I have no idea about whether the jewelry is genuine or not.

Mary：Please set your heart at rest, we can show you their certificates of authority.

Kate：Are those tourmalines are natural or treated?

Mary：They have already been heated equally to natural, permanent beauty.

Kate：I had heard about radiated tourmaline, tell me about it?

Mary：That is the way to produce a new color, unequally to natural. We have no radiated tourmaline all the time.

Kate：I am confused about equally or unequally to natural.

Mary：The acceptable treated gemstones way is equal to natural; whereas unacceptable treated gemstones way is unequal to natural. By the way, there are some rules about it.

Kate：Now, I see, thank you.

Mary：That's all right, everybody is welcome here, whether she buys or not, please take your time.

Kate：Thank you very much.

Part 2 Fill the Chinese Meanings with the Teacher's Tutorship

inclusion _____   halo _____   flaw _____   gas bubble _____
fingerprint _____   negative crystal _____   internal world of gemstone _____
unaided eye _____   pre-existing inclusion _____   secondary inclusion _____

Part 3 Link the Relevant Words between Column A and Column B

Column A	Column B
rutile	glass
curved striae	cat's-eye
gas bubbles	YAG
yttrium aluminum garnet	quartz
radiating horsetail inclusion	demantoid
parallel inclusion	flame fusion synthetic ruby

Part 4 Useful Phrases

cut, set and polish gem stones 切割、镶嵌、磨光宝石
inclusions of mineral grains in natural garnet 天然石榴石中的矿物包裹体
silk inclusions in natural ruby 天然红宝石中的丝状包裹体
quality, characteristics and value of gem stones 宝石的质量、性质和价值
zircon halo in natural ruby 天然红宝石中的锆石晕
three phase inclusions in natural emerald 天然祖母绿中的三相包裹体
gas bubbles & gas clouds in synthetic ruby 合成红宝石中的气泡和云状物
veil-like inclusions 面纱状包裹体
tone, brightness saturation of color 颜色的色调、明度和饱和度
primary inclusion 原生包裹体
secondary inclusion 次生包裹体
yttrium aluminum garnet(YAG) 钇铝榴石
honeycomb structure 蜂窝状结构
bread-crumb like inclusion 面包渣状包裹体
infrared spectroscope 红外光谱仪
raman spectrometer 拉曼光谱仪
Gemological Institute of America(GIA) 美国宝石学院

Part 5 Some Sentences

(1) The characteristics of inclusions in natural and synthetic gem are distinct owing to

their different forming mechanisms and can be utilized to distinguish gems from different localities.

（2）Gem enhancements such as surface diffusion, bulk diffusion, beryllium treatments, etc. can only be detected with the stone immersed. Amateurs use water and baby oil in an immersion dish; professionals use a liquid which is close to the refractive index of the stone to be tested.

（3）Inclusions provide invaluable information about the identity of their gem host, sometimes revealing a stone's origin, sometimes betraying alteration of a natural stone through treatment, and sometimes unmasking a synthetic or simulant masquerading as a natural stone.

Part 6 Short Paragraph

Example of Inclusions in Diamond and Emerald

Diamond inclusions are characteristics that occur inside a stone. They are usually called flaws, because their presence means the diamond is not perfect.

Inclusions are like fingerprints, a characteristic that gives us all a special signature. Getting to know your diamond inside and out makes the stone a more personal possession and will help you describe and identify the gem if it is ever lost or stolen.

Some inclusions affect a diamond's clarity, making it less brilliant because they interfere with light as it passes through the stone. Other types of diamond inclusions can make a gemstone vulnerable to shattering.

There are few perfect diamonds, and the ones that are perfect are quite expensive, so the diamonds we buy all have varying amounts of internal and external flaws. Most jewelers tell us not to worry about diamond inclusions if they do not affect the stone's strength or seriously impact its appearance.

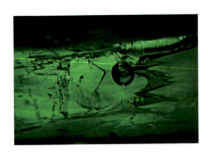

Almost all natural emeralds contain inclusions and these are very important in distinguishing natural from synthetic emeralds and other green stones. Some inclusions are typical for a particular locality. This, when combined with an assessment of RI and SG, may provide an indication of the country of origin.

Look at the picture above. Most emeralds contain numerous fractures and openings. This photo shows the classic three-phase inclusion of a Colombian emerald, consisting of a negative crystal filled with a gas bubble, a halite crystal and a brine solution. If cutting exposes this inclusion to the surface, it will fill with air, producing reflection. Such inclusions are typically filled with oil or resin to reduce the amount of reflection.

Part 7 Text

Gemstone Inclusion

What is an inclusion? John Koivula (1991) has provided us with probably the best definition of the word "inclusion": broadly defined, an inclusion is any irregularity observable in a gem-by the unaided eye or ⌈using⌋ some tool such as a hand lens or microscope. The "irregularity" may be a substance, such as a solid mineral crystal or a fluid filling a cavity, or it may be an unfilled cavity, a fracture, or a growth pattern that produces some optical effect.

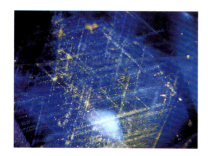

We can use the internal characteristics of gemstones to extrapolate the conditions in which they were formed, including temperature and pressure, and how that relates to the geological and geographical region they may have been associated with. The original locality of a particular gemstone can be traced back to the source by diagnostic features like a specific solid crystal inclusion or the presence of a three-phase inclusion, just to name a couple.

The internal world of gemstones reveals mineralogical clues to trace a gem's origin and what has happened to it on its way from the mine to the jeweler. Enhancements and treatments are even more of a factor today since the average gem collector has the infinite resources of the internet to research and explore the secrets of the gem trade. Understanding internal characteristics can help identify the presence of heat treatment and distinguish a natural from a synthetic.

A gemstone may also vary from the normal physical property values due to inclusions which can, if present in sufficient quantities, reduce or increase the specific gravity. This is true of many gemstones, so allowances must be made when assessing the SG of any particular specimen, since most printed information will probably be based upon good

quality, inclusion-free material, some gemstones rely upon large numbers of inclusions for visual appeal, as with star, cat's eye and aventurine stones.

Inclusions in gemstones can be classified according to the scheme proposed by Gübelin (1973) and Gübelin & Koivula (1986), which is based upon their age with respect to that of the host crystal. This is as follows.

Pre-existing Inclusions (protogenetic)

Inclusions that have formed before the host. These are strictly of a solid nature (pre-existing liquids and gases don't count).

Solid and semi-solid inclusions: crystals and/or glasses that form before the host and are subsequently trapped. The crystals may appear either as heavily etched or corroded individuals which formed long before the host, or as well-formed crystals which developed just prior to the host. Examples: various, including spinel in ruby. Corundums which formed in metamorphic environments, such as Burmese rubies, are often rich in solid inclusions.

Contemporary Inclusions (syngeneic)

Inclusions that have formed at the same time as the host.

Secondary Inclusions (epigenetic)

Inclusions that have formed immediately, or even millions of years, after the host stopped growing.

Please look at following figures and learn the corresponding explanations.

(1) Spinel octahedra in a blue sapphire from Sri Lanka; 25×.

(2) Iridescent two-phase primary inclusions in Thai/Cambodian ruby, parallel to the basal plane; 25×.

(3) Primary rutile crystals (orange) in a sapphire from Rock Creek, Montana; 25×.

(4) Secondary, exsolved rutile "silk" clouds in a sapphire from Rock Creek, Montana; 20×.

(5) Light yellow apatite crystals in a corundum from Umba Valley, Tanzania; 25×.

(6) Discoid fractures caused by heat treatment in an Australian sapphire; 45×.

(7) Solid inclusions in a Sri Lankan sapphire, viewed between crossed polaroids. Note the strain and interference colors on the crystals.

(8) A primary negative crystal in a Sri Lankan sapphire. The cavity contains liquid and gaseous CO_2, along with a mobile graphite crystal cluster; 35×.

(9) Terraced basal pinacoid faces on Burmese ruby crystals. Note the many tiny triangular etch (or growth?) marks. Such stepped surfaces result from oscillatory growth between the basal pinacoid and prism or pyramid/rhombohedron faces and are often seen on ruby crystals, particularly those from Burma.

(10) Angular growth zoning in sapphires from Australia. Such zoning may be found parallel to any of the crystal faces. It is never curved if viewed exactly parallel to the crystal

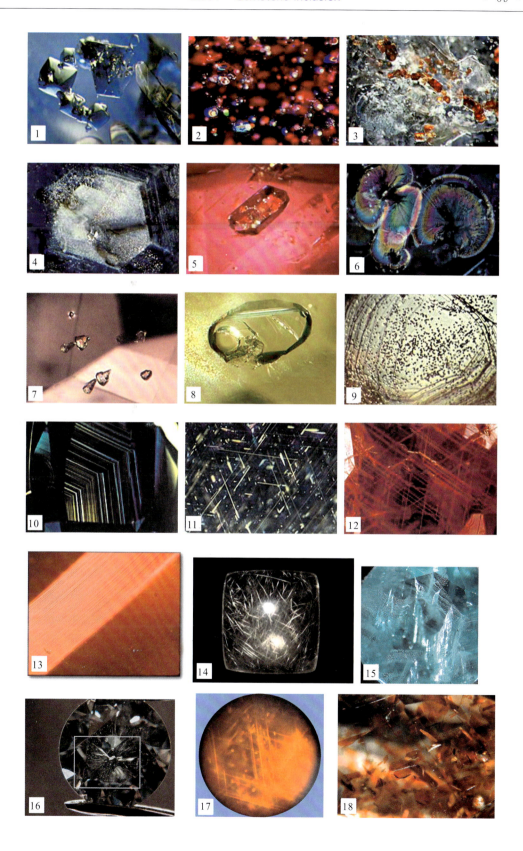

face along which it formed.

(11) A magnified view of the silk. Note the arrow shapes of the rutile silk. Such perfectly formed silk is proof that a specimen has not been subjected to high-temperature heat treatment.

(12) Boehmite needles in corundum are commonly confused with rutile silk, but are easily separated by their orientation. Exsolved boehmite unmixes in three directions parallel to the edges of the rhombohedron (three directions in total, but only two in the same plane. The needles intersect at 86.1°/93.9°. Boehmite needles are typically found at the junctions of crossing twin planes).

(13) Curved striae clearly visible in this flame fusion synthetic ruby.

(14) This genuine golden rutilated quartz gemstone was finished and polished from natural rutile in quartz crystal rough that was mined in Brazil. This natural untreated golden rutile gemstone has beautiful gold rutile needles passing through very clear rock crystal quartz.

(15) Flux "fingerprints" are repaired remannant of a fracture in a Synthetic Emerald.

(16) Unidentified crystal is in the center of the horsetail nebula in demantoid. From a 2.81 carat round, 8.0mm×5.4mm.

(17) The needle-like inclusion in garnet.

(18) Randomly oriented hematite platelets exhibit interference colors under oblique illumination. The host is a very unusual sunstone from Tanzania.

Part 8 Words and Expressions

recommend[ˌrekəˈmend]	vt. 推荐,介绍
genuine[ˈdʒenjuin]	adj. 真实的,真正的,诚恳的
certificate[səˈtifikit]	n. 证书,证明书
authority[ɔːˈθɔriti]	n. 权威
radiated[ˈreidieitid]	adj. 辐射的
confounded[kənˈfaundid]	adj. 糊涂的,困惑的,讨厌的(轻微的诅咒用语)
inclusion[inˈkluːʒən]	n. 内含物,包裹体,包含
halo[ˈheiləu]	n. 晕,晕圈
flaw[flɔː]	n. 瑕疵,缺点
gas bubble	气泡
fingerprint[ˈfiŋɡəprint]	n. 指纹,手印
negative crystal	负晶
internal world of gemstones	宝石内含物
pre-existing inclusions	原生包裹体
striae[ˈstraiə]	n. 条纹,细沟
curved striae	弯曲的生长纹
yttrium[ˈitriəm]	n. 钇(稀有金属元素,符号 Y)
horsetail[ˈhɔːsteil]	n. 马尾

bulk[bʌlk]	n. 大小, 体积, 大批
bulk diffusion	体扩散
invaluable[ɪnˈvæljuəbl]	adj. 无价的, 价值无法衡量的
betray[bɪˈtreɪ]	vt. 出卖, 背叛, 泄露(秘密), 露出……迹象
unmasking[ˌʌnˈmɑːskɪŋ]	vi. 脱去假面具
masquerading[ˌmɑːskəˈreɪdɪŋ]	n. 化装舞会 v. 化装
brilliant[ˈbrɪljənt]	adj. 灿烂的, 闪耀的, 有才气的
interfere[ˌɪntəˈfɪə]	vi. 干涉, 干预, 妨碍, 打扰
vulnerable[ˈvʌlnərəbl]	adj. 易受攻击的, 易受……的攻击
shatter[ˈʃætə]	n. 粉碎, 碎片, 落花(叶, 粒等) vt. 打碎, 使散开, 粉碎 vi. 粉碎, 损坏, 落叶
impact[ˈɪmpækt]	n. 碰撞, 冲突, 影响, 效果
particular[pəˈtɪkjulə]	adj. 特殊的, 特别的
Colombian[kəˈlɒmbɪən]	n. 哥伦比亚人
halite[ˈhælaɪt]	n. 岩盐
brine solution	盐溶液
irregularity[ɪˌregjuˈlærɪtɪ]	n. 不规则, 无规律
fluid[ˈfluːɪd]	n. 流动性, 流度
extrapolate[eksˈtræpəleɪt]	v. 推断, [数]外推
diagnostic[ˌdaɪəɡˈnɒstɪk]	adj. 诊断的
mineralogical[ˌmɪnərəˈlɒdʒɪkəl]	adj. 矿物学的
infinite[ˈɪnfɪnət]	n. 无限的东西(如空间、时间), [数]无穷大
allowance[əˈlauəns]	n. 津贴, 补助, 宽容, 允许 vt. 定量供应
appeal[əˈpiːl]	n. 请求, 呼吁, 上诉, 吸引力, 要求
aventurine[əˈventʃərɪn]	n. 砂金石
scheme[skiːm]	n. 安排, 配置, 计划, 阴谋, 方案, 图解, 摘要 v. 计划, 设计, 图谋, 策划
protogenetic[ˌprəʊtədʒɪˈnetɪk]	adj. 原生的
host[həʊst]	n. 主人, 旅馆招待, 许多 vt. 当主人招待, 主机
subsequently[ˈsʌbsɪkwəntlɪ]	adv. 后来, 随后
corrode[kəˈrəʊd]	v. 使腐蚀, 侵蚀
prior to	adv. 在前, 居先
various[ˈveərɪəs]	adj. 不同的, 各种各样的, 多方面的, 多样的
metamorphic[ˌmetəˈmɔːfɪk]	adj. 变质的
Burmese[bɜːˈmiːz]	n. 缅甸人, 缅甸语 adj. 缅甸的, 缅甸人的, 缅甸语的
contemporary[kənˈtempərərɪ]	n. 同时代的人 adj. 当代的, 同时代的
syngeneic[ˌsɪndʒɪˈniːɪk]	adj. 同源的, 同基因的, 同系的
octahedra[ˌɒktəˈhedrə]	n. 八面体
Sri Lanka[srɪˈlæŋkə]	n. 斯里兰卡(南亚岛国)

iridescent[ˌɪrɪˈdesnt]	adj. 彩虹色的,闪光的
Montana[mɒnˈtænə]	n. 蒙大拿(美国州名)
Tanzania[ˌtænzəˈniːə]	n. 坦桑尼亚(东非国家)
discoid[ˈdɪskɔɪd]	adj. 铁饼状的
strain[streɪn]	n. 过度的疲劳,紧张,应变 vt. 扭伤,损伤 v. 拉紧,扯紧,(使)紧张,尽力
graphite[ˈɡræfaɪt]	n. 石墨
terraced[ˈterəsɪd]	adj. 台地的,阶地的
pinacoid[ˈpɪnəkɔɪd]	n. 轴面
triangular etch[traɪˈæŋɡjulə][etʃ]	三角痕
oscillatory[ˈɒsɪleɪtəri]	adj. 摆动的
prism[ˈprɪzəm]	n. 棱镜,棱柱
rhombohedron[ˌrɒmbəˈhedrən]	n. 菱面体
specimen[ˈspesɪmən]	n. 范例,标本,样品,样本,待试验物
oblique[əˈbliːk]	adj. 倾斜的,间接的,不坦率的,无诚意的
boehmite[ˈbɔːmaɪt]	n. 勃姆石(一水软铝石)
Brazil[brəˈzɪl]	n. 巴西
incandescent light[ˌɪnkænˈdesnt][laɪt]	白炽光
nebula[ˈnebjulə]	n. 星云,云翳

Part 9 Check Your Understanding

Exercise 1 Answer the following questions

(1) What is the inclusion of gemstone?
(2) Why do the gemmologists study on the inclusions of gemstones?
(3) Which gemstones are routinely heated treatment?
(4) What is the secondary inclusions?
(5) How can we see the inclusions of gemstones?

Exercise 2 Fill in the blanks

(1) Please illustrate several appearances of inclusions in some gemstones: _____, _____, _____, _____.
(2) Some inclusions affect a diamond's _____.
(3) Primary _____ crystals in ruby and sapphire is familiar.
(4) When a large numbers of parallel inclusions is in some gemstones, there will be peculiar optical phenomena, such as _____ and _____.

Exercise 3 Translation

(1) The characteristics of inclusions in natural and synthetic gem are distinct owing to their different forming mechanisms.
(2) There are few perfect diamonds, and the ones that are perfect are quite expensive, so the

diamonds we buy all have varying amounts of internal and external flaws.

(3) A gemstone may also vary from the normal physical property values due to inclusions which can, if present insufficient quantities, reduce or increase the specific gravity.

(4) Inclusions in gemstones can be classified into pre-existing inclusions, contemporary inclusions and secondary inclusions.

(5) Laser drilling removes inclusions, improving clarity.

(6) 哥伦比亚祖母绿中可能出现三相包裹体。

(7) 我们可以通过放大观察包裹体来鉴别合成宝石和优化处理宝石。

(8) 在焰熔法合成的蓝宝石中可能见到弧形生长纹包裹体。

Part 10 Self-Study Material

Types of Diamond Inclusions

Crystals and Mineral Inclusions

Diamonds can have tiny crystals and minerals imbedded in them, even other diamonds. Many cannot be seen without magnification, but a large chunk or grouping of crystals that detracts from a diamond's appearance lowers its clarity grade, and its value.

There are times that a small crystal can add character to a diamond. A diamond with a small garnet imbedded in it would be a great conversation piece and a personal choice for someone whose birthstone is a garnet.

Pinpoint Inclusions

Pinpoints are tiny light or dark crystals in diamonds that appear by themselves or in clusters. Larger clusters of minute pinpoints can create a hazy area in the diamond, called a cloud, which affects the diamond's clarity.

Laser Lines

Laser lines are not a natural diamond inclusion. These vapor-like trails are left behind when lasers are used to remove dark inclusions from the diamond. The machine-made trails look like tiny strands of thread that begin at the diamond's surface and stretch inward, stopping at the point where the inclusion was removed.

Feathers

Feathers are cracks within the stone that resemble, well……feathers. Small feathers do not usually affect a diamond's durability unless they reach the surface on the top of the stone, a location that's prone to accidental blows.

Cleavage

Diamond cleavage is a straight crack with no feathering. A cleavage has the potential to split the diamond apart along its length if it is hit at the correct angle.

Small cracks that are not visible when a diamond is viewed in a table-up (face up) position do not seriously affect clarity ratings.

Girdle Fringes, Bearding

Girdle fringes, or bearding, are hair-like lines that can occur around the girdle during the cutting process. Minimal bearding is usually not a problem, but extensive fringing is often polished away or removed by recutting the diamond.

Grain Lines, Growth Lines

Grain lines are created by irregular crystallization that takes place when a diamond is formed. Colorless grain lines do not usually affect diamond clarity unless they are present in large masses. White or colored grain lines can lower a diamond's clarity grade.

Always shop for diamonds at a jewelry store you trust and find someone who can answer your questions about the diamonds you are considering. Ask the jeweler to show you each diamond under magnification and explain its characteristics to you.

Read *Surface Blemishes on Diamonds* for more information about flaws that can affect diamond clarity and strength.

Unit 8　Gemstones Identification Procedure

Part 1　Dialogue

Mary: Morning. Can I help you?
Kate: I want a first-rate loose ruby.
Mary: What shape do you prefer? Round? Pear? Oval? Square? Heart?
Kate: I'm ready to make a pendant with 18K-gold. What do you think?
Mary: Oh, pear-shape is the best, I think.
Kate: Could you choose one for me?
Mary: I recommend this one, madam. The ruby with vivid red and moderate grain size.
Kate: Thanks. Where is the ruby from?
Mary: Burma. With high quality and top-polish.
Kate: How much does it cost?

本章音频

Mary: We sell it by carats. Wait a minute, please.
Kate: Sure.
Mary: 0.88 of a carat, ＄720.
Kate: Can you make it cheaper?
Mary: I'll give it to you for ＄720, ok?
Kate: We have to ask for another price reduction.
Mary: Well, ＄680, you can think about another cut.
Kate: I see, thanks a lot.

Part 2 Fill the Chinese Meanings with the Teacher's Tutorship

identify _____ characteristic _____ artificial _____ lens _____
unset _____ dichroism _____ pleochroism _____ trichroism _____
birefringence _____ vitreous _____ distinguish _____ substitute _____

Part 3 Link the Words and the Relevant Pictures

RI(refractive index)   SG(specific gravity)   Dis(dispersion)

diamond

spinel

tourmaline

sapphire

emerald

chalcedony

1. RI:1.718, single Pleochroism:None
2. RI:1.624~1.644
   SG:3.05
   Luster:vitreous
3. RI:1.577~1.583
   SG:2.72
   Luster:vitreous uniaxial(—)
4. RI:1.762~1.770
   SG:4.00
   Luster:strong vitreous
5. Luster:adamantine
   Dis:0.044
6. RI:1.54±
   SG:2.65
   Luster:weak vitreous

Part 4 Useful Phrases

sterling silver   标准纯银，纯银
tiger's-eye   虎睛石
alexandrite cat's-eye   变石猫眼
blood jasper   血玉髓
bowenite(serpentine jade)   鲍文玉（硬绿蛇纹石）
cluster setting   群镶
flame opal   火欧泊
flint glass   铅玻璃

Kashmir sapphire   克什米尔蓝宝石
marriage ring   结婚戒指
reconstructed amber   再造琥珀
scenic agate   风景玛瑙
natural organic substances   天然有机宝石
artificial products   人工宝石
composite stones   拼合宝石

Part 5 Some Sentences

(1) Stones can be dyed, heated or irradiated in an attempt to make them more attractive. Some stones are treated to fill in cracks. Others, like emeralds, may have been altered with the addition of a layer of synthetic gem.

(2) Is there a difference in luster between crown and pavilion, or between different portions of the crown? Such differences are common in garnet and glass and other doublets or triplets with wide differences in refractive index between the parts.

(3) Luster can be further characterized as metallic, sub-metallic, adamantine, sub-adamantine, vitreous, sub-vitreous, waxy, greasy, silky, or dull. The first three reflect the presence of refractive indices over the refractometer scale.

(4) Jewelers typically do the handwork required to produce a piece of jewelry, while gemologists study the quality, characteristics and value of gem stones. Gemologists usually sell jewelry and provide appraisal services.

Part 6 Short Paragraph

Some Hints in Identifying Gemstone

Some hints could remind you of identifying gemstones when you observed the counterpart.

If any of the various optical phenomena are present—play of color, change of color and adularescence—the number of possibilities is reduced materially, weak asterism and chatoyancy are found in a number of species. Asterism is frequently seen in ruby, sapphire and orthoclase. Stones that show a cat's eye effect include the familiar chrysoberyl, quartz and tourmaline, but so do beryl, demantoid, nephrite, enstatite, diopside, feldspars, apatite, zircon, sillimanite and others.

A transparent, faceted stone that shows a red ring near the girdle when it is turned table-down on a white surface suggests a garnet-topped doublet. Flashes of red from a deep,

vivid blue stone suggest synthetic spinel, or the tanzanite variety of zoisite.

Both zircon and synthetic rutile have exceedingly high birefringence, a condition easily recognized in transparent materials under magnification.

If there is strong doubling of opposite facet edges and the stone has natural inclusions, synthetic rutile is eliminated, and the unknown must be zircon. A doublet could be detected under magnification, or if there are bubbles and no doubling, glass. If the stone proved to be a diamond, however, only the spectroscope could distinguish between naturally and artificially colored material.

If the stone is doubly refractive with no visible inclusions, then specific gravity, strength of doubling, or strength of dispersion could distinguish between high-property zircon and synthetic rutile. Immersion in methylene iodide would show a great difference in refractive index by a great difference in relief.

Part 7 Text

Gemstones Identification Procedure

Those familiar with the appearance of the important gemstone species can usually narrow the identity of an unknown stone down to a few possibilities. For example, an unknown is set in jewelry, it is unlikely to be one of these species cut almost exclusively for collectors. In such a case, the total number of likely possibilities is reduced to about 25. The hue, tone and saturation of the color reduce the number of possible species still further.

In some colors, there are only a few possibilities for a non-transparent stone, for example, a banded stone in two shades of green is almost sure to be malachite, dyed agate, dyed onyx marble, glass, or plastic. Transparent colored and colorless stones suggest several possibilities, but these can be reduced either slightly or materially, depending on the hue and tone. A practiced glance, noting only color and transparency, has a distinct value and enables an observant gem tester to reduce the list to a very small number.

The nature of the color is important. Is it light, medium,

Unit 8　Gemstones Identification Procedure

or dark? Is its intensity vivid or dull? Spodumene, for example, is never dark in tone. Only a few gem materials occur in a vivid chrome green: emerald, jadeite, demantoid garnet, tourmaline, diopside, dioptase and dyed or backed stones. These and other immediate observations are important.

When a colored stone is turned, the tester should look for any obvious pleochroism. Common gemstones with sufficient pleochroism to be noted with the unaided eye include kunzite, andalusite, tourmaline, zircon, ruby, sapphire and alexandrite. Among rarer stones, pleochroism can be obvious in kornerupine, iolite, epidote, tanzanite and a few others.

Another important characteristic in the initial examination is the luster. Since this is determined by the refractive index (along with the flatness of the polished surface), the higher the luster, the higher the refractive index of the unknown. Sub-adamantine suggests an index high on the scale; vitreous, mid-scale; and sub-vitreous, low. Waxy and greasy luster are usually associated with poorly polished surfaces. While silky refers to stones with many needle-like inclusions. Comparison with gems of known identity helps classify refractive indices readily.

With translucent and opaque materials, the luster on fracture surfaces is particularly important. Most transparent stones in the middle to low refractive index range have a vitreous luster on conchoidal fracture surfaces, as do glass imitations. Many natural translucent and opaque stones, however, have granular or other types of fracture. Those with conchoidal fractures seldom have a vitreous luster. Chalcedony usually has a waxy luster on fractures; turquoise is dull. This is a ready means of separating natural stones from glass with its vitreous fracture luster.

The degree to which dispersion is evident in a transparent, faceted stone provides another important clue to its identity. Only a few gemstones and their substitutes have sufficient dispersion to be obvious and noteworthy to the unaided eye. Stones strong in dispersion included synthetic rutile, strontium titanate, demantoid, sphene, diamond, benitoite, zircon and some glass. The presence or absence of fire is significant.

Is doubling of opposite face junctions obvious to the unaided eye or under low magnification? Among important gem materials, stones the show strong doubling include synthetic rutile, sphene, zircon, peridot and tourmaline. Similarly,

only a few species, such as diamond, topaz, spodumene and the feldspars, display obvious cleavage.

How well the stone is polished may suggest its hardness range. Stones with rounded facet edges and poor polish are generally soft. On the other hand, Synthetic corundum and other inexpensive materials are sometimes polished so rapidly that the polish is inferior. On the surface of synthetic corundum irregular fractures caused by the heat generated in too rapid polishing are typical.

Other characteristics assist in narrowing the number of possibilities of an unknown as well. After noting any that are obvious to the unaided eye or a low-power loupe, the next step is to determine the refractive index or indices. If it is done in monochromatic light or with a filter, this often eliminates all but one stone. At other times, only two or three possibilities exist.

From this point on, the procedure depends on the refractometer findings. If the refractive index by itself is insufficient to identify an unknown, the next step is to separate the remaining possibilities. If, for example, the refractive index of a transparent yellow gem is above the limits of the refractometer, there are several possibilities. Diamond, synthetic cubic zirconia, zircon, or synthetic rutile is most likely, but diamond could be colored naturally, or by irradiation and heat. A diamond doublet is also a

possibility and although it is unlikely, glass too could have an index above the limits of the refractometer, as could the garnet top of a garnet and glass doublet.

Taking our example further, several things must be determined, with each successive step indicated by the results of the previous test. Perhaps the first information needed is whether the unknown is singly or doubly refractive. This could be determined with a polariscope, but since we also need to know some characteristics as seen under magnification, the use of a magnifier is the next logical step. The number of tests depends on the findings of the test that went before. Thus our unknown yellow transparent stone might call for one or two tests, or a half dozen.

Part 8　Words and Expressions

first-rate　　　　　　　　　　　　adj. 最上等的
loose ruby　　　　　　　　　　　未镶嵌的红宝石
pear shape　　　　　　　　　　　梨形
vivid['vɪvɪd]　　　　　　　　　　adj. 鲜明的

artificial[ˌɑːtɪˈfɪʃəl]	adj. 人造的
unset[ˌʌnˈset]	adj. 未凝固的；未安装的
pleochroism[plɪˈɔkrəuɪzəm]	n. 多色性
trichroism[ˈtraɪkrəuɪzəm]	n. 三色性
birefringence[ˌbaɪrɪˈfrɪndʒɪns]	n. 双折射
vitreous[ˈvɪtrɪəs]	adj. 玻璃质的
distinguish[dɪsˈtɪŋgwɪʃ]	v. 区别；辨别
substitute[ˈsʌbstɪtjuːt]	n. 代用品；代替者；替代品　v. 代替；替换；替代
RI (refractive index)	折射率
SG (specific gravity)	比重（相对密度）
Dis (dispersion)	色散
uniaxial[ˌjuːnɪˈæksɪəl]	adj. 单轴的；单轴晶体
adamantine[ˌædəˈmæntaɪn]	adj. 金刚石的
irradiate[ɪˈreɪdɪeɪt]	v. 照射
crown[kraun]	n. 王冠；花冠；顶
sub-metallic[ˌsʌbmɪˈtælɪk]	adj. 亚金属的
waxy[ˈwæksɪ]	adj. 蜡质的
greasy[ˈgriːzɪ]	adj. 油脂的
dull[dʌl]	adj. 阴暗的
reflect[rɪˈflekt]	v. 反射
refractometer[ˌriːfrækˈtɔmɪtə]	n. 折射仪
hint[hɪnt]	n. 暗示
semblable[ˈsembləbl]	adj. 相似的；外表（上）的
play of color	变彩效应
change of color	变色效应
adularescence[ˌædjuləˈresəns]	冰长石晕彩
materially[məˈtɪərɪəlɪ]	adv. 本质上；物质上；重大
chatoyancy[ʃəˈtɔɪənsɪ]	n. 猫眼效应
orthoclase[ˈɔːθəukleɪs]	n. 正长石
enstatite[ˈenstətaɪt]	n. 顽火辉石
diopside[daɪˈɔpsaɪd]	n. 透辉石
sillimanite[ˈsɪlɪmənaɪt]	n. 矽线石
transparent[trænsˈpærənt]	adj. 透明的
girdle[ˈgɜːdl]	n. 腰
zoisite[ˈzɔɪsaɪt]	n. 黝帘石
procedure[prəˈsiːdʒə]	n. 程序；手续
marble[ˈmɑːbl]	n. 大理石
spodumene[ˈspɔdjumiːn]	n. 锂辉石
dioptase[daɪˈɔpteɪs]	n. 绿铜矿

kunzite[ˈkuntsaɪt]　　　　　　　　　　　　n. 紫锂辉石
andalusite[ˌændəˈluːsaɪt]　　　　　　　　n. 红柱石
kornerupine[ˌkɔːnəˈruːpin]　　　　　　　n. 柱晶石
epidote[ˈepɪdəut]　　　　　　　　　　　n. 绿帘石
fracture[ˈfræktʃə]　　　　　　　　　　　n. 破裂
granular[ˈgrænjulə]　　　　　　　　　　adj. 由小粒而成的；粒状的
sphene[sfiːn]　　　　　　　　　　　　　n. 楣石
benitoite[bəˈniːtəuaɪt]　　　　　　　　　n. 蓝锥矿
junction[ˈdʒʌŋkʃn]　　　　　　　　　　n. 连接；接合
inferior[inˈfɪərɪə]　　　　　　　　　　　adj. 劣质的
monochromatic[ˌmɔnəukrəuˈmætɪk]　　adj. 单色的

Part 9　Check Your Understanding

Exercise 1　Answer the following questions

(1) How do you check the doublets or triplets with unaided eye?
(2) How do you distinguish zircon from synthetic rutile? They are both with exceedingly high birefringence.
(3) How do you distinguish between ruby and red tourmaline?
(4) What species stones are with strong dispersion?
(5) What clues were revealed by the hardness in identifying gemstones?

Exercise 2　Fill in the blanks

(1) The luster of sapphire is _____; The luster of Hetian Yu is _____.
(2) Play of color is often appeared on _____ (gemstone's name).
(3) Please illustrate several commonly nontransparent stones：_____.
(4) Please illustrate several commonly vivid chrome green stones：_____.

Exercise 3　Translation

(1) If any of the various optical phenomena are present—play of color, change of color, and adularescence—the number of possibilities is reduced materially, weak asterism and chatoyancy are found in a number of species.
(2) In some colors, there are only a few possibilities for a nontransparent stone—for example, a banded stone in two shades of green is almost sure to be malachite, dyed agate, dyed onyx marble, glass, or plastic.
(3) How well the stone is polished may suggest its hardness range. Stones with rounded facet edges and poor polish are generally soft.
(4) 鉴定宝石时，可以先肉眼观察宝石的颜色、光泽、透明度等特点。
(5) 折射仪和天平是鉴定宝石的重要仪器。
(6) 我们不仅要鉴别出宝石的种属，还要知道它是天然的还是合成的，是否经过人工优化或者处理。

Part 10 Self-Study Material

Characteristics of Some Stones

Ruby

chemical formula	Al_2O_3
crystal habit	massive and granular
crystal system	hexagonal
cleavage	none
fracture	conchoidal, splintery
hardness(Mohs' scale)	9.0
luster	vitreous
refractive index	1.762~1.778
optical properties	uniaxial(−)
birefringence	0.008
pleochroism	strong
specific gravity	3.95~4.03
melting point	2030~2050℃

Ruby is from the corundum family. It is red because of the presence of the trace element chromium. Ruby naturally occur in nature and it is common to heat (thermal enhancement) them to improve their color. The untreated ones are rare and command a premium.

Emerald

Emerald is from the beryl family. It is green because of the presence of the element chromium. Emeralds naturally occur in nature and it is common to oil them to improve their clarity and this is a general accepted practice. The untreated ones are very rare and command a premium. Emeralds originate from various origins such as Columbia, Brazil, Zimbabwe.

chemical formula	$Be_3Al_2[Si_6O_{18}]$
crystal habit	hexagonal crystals
crystal system	hexagonal
cleavage	poor basal cleavage
fracture	conchoidal
hardness(Mohs' scale)	7.5~8.0
luster	vitreous
refractive index	1.576~1.582
pleochroism	distinct, blue-green/yellow-green
streak	white
specific gravity	2.70~2.78

Citrine

Citrine comes from the Quartz family. It gets its violet because of the presence of the trace element iron and aluminium. Citrines naturally occur in nature but might be heated from amethyst or smoky quartz. Major source of citrine is Brazil.

chemical formula	SiO_2
crystal habit	6 sided prism ending in 6 sided pyramid
crystal system	trigonal
cleavage	none
fracture	conchoidal
hardness (Mohs' scale)	7
luster	vitreous
refractive index	1.544~1.553
optical properties	uniaxial(+)
birefringence	+0.009
pleochroism	none
specific gravity	2.65
other characteristics	piezoelectric

Unit 9　Gemstone Cut

Part 1　Dialogue

本章音频

Mary：Are you being helped?

Kate：Do you have peridot rings, heart shape, size 17?

Mary：You are great, to learn the gem-cut more, oval-cut is ok?

Kate：No, I have seen my friend's ring like heart, I want a similar one.

Mary：Come over, you can find more beautiful jewelry possibly.

Kate：All kinds of cut! I can't get out from the loveliness.

Kate：Valentine's day is coming, I want to buy some jewelry for my boy friend.

Mary：Here is a set of sweetie-silver ring, from Italy, excellent handwork.

Kate：Ok, that's a good idea, let me think about it.

Mary：Don't hesitate, lady. In fact, it's the latest fashion. This style ring is our best sale recently.

Kate：It is reasonable. Is it of the highest quality?

Mary：Sure, the craft is first-rate. You'd better try it now.

Kate：Yes. But how much is it?

Mary：$880.

Kate：Can you give me a discount?

Mary：The whole set is cheaper. We have offered a ten percent discount. We can't reduce the price any further.

Kate：In any case, it's a nice set, I'll take it.

Part 2 Fill the Chinese Meanings with the Teacher's Tutorship

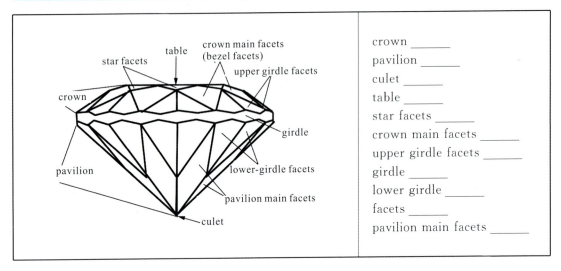

crown _____
pavilion _____
culet _____
table _____
star facets _____
crown main facets _____
upper girdle facets _____
girdle _____
lower girdle _____
facets _____
pavilion main facets _____

Part 3 Link the Words and the Relevant Pictures

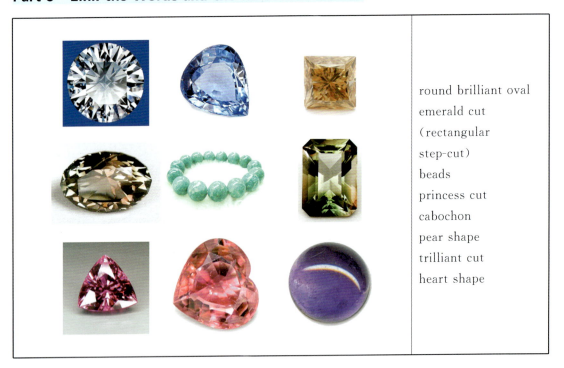

round brilliant oval
emerald cut
(rectangular
step-cut)
beads
princess cut
cabochon
pear shape
trilliant cut
heart shape

Part 4 Useful Phrases

facet cut   翻光面琢型
professional jewelcrafter   专业的宝石加工师
double cabochon   双弧面型
fancy cut   异形切工

brilliant step cut   多面阶梯琢型
rough stone   原石
Lisbon cut   里斯本琢型
gem studded ring   镶宝石指环
cutting style   琢型
cushion cut   长角阶梯切割
marquise-cut diamond   榄尖形切割
channel set   槽镶
pave set   群镶
a hand lens   手持放大镜
false amethyst   假紫晶
fancy agate   奇珍玛瑙
Japan-law twin-like quartz   日本双晶的石英

Part 5   Some Sentences

(1) A faceted gem cutting process begins with the selection of faceting rough, which is chosen for color, clarity, type of stone, shape and perhaps some other qualities as well.

(2) Carving is the most challenging of the lapidary arts and there are very few recognized experts in the field. To be successful, one must have a distinctive artistic sense and a thorough understanding of the principles of lapidary. Unlike working in wood or metal, the materials present definite limits as to what can be done.

(3) Composit gemstones are created by bonding two or more materials together to form one gemstone. When the final gem is made of two different materials, it is called a "doublet", and when made with three, a "triplet". Common doublet or triplet gemstones are emerald and opal, usually combined with quartz or glass. Compositing is used to improve color and add durability. In faceted composite stones, the crown would be the natural material.

Part 6   Short Paragraph

Gemstone Cut Shapes

Round

This shape has set the standard for all other diamond shapes and accounts for more than 75% of diamonds sold today. Its 58-facet cut is calibrated through a precise formula to achieve the maximum in fire and brilliance.

Princess

This is a square or rectangular cut with numerous sparkling facets. It is a relatively new cut and often finds its way into solitaire engagement rings. Flattering to a hand with long fingers, it is often embellished with triangular stones at its sides.

Emerald

This is a rectangular shape with cut corners. It is known as a step cut because its concentric broad, flat planes resemble stair steps.

Oval

This is an even, perfectly symmetrical design popular among women with small hands or short fingers. Its elongated shape gives a flattering illusion of length to the hand.

Pear

This hybrid cut, combining the best of the oval and the marquise, is shaped most like a sparkling teardrop. It also belongs to that category of diamond whose design most complements a hand with small or average-length fingers. It is particularly beautiful for pendants or earrings.

Cushion

The cushion-cut is a deep cut with large facets, an open bottom and rounded corners. The depth of this cut encourages the eye to travel around the inside the stone. The facets create a high degree of returning light, making the cushion-cut one of the most brilliant and sparkling diamond cuts.

Heart

This ultimate symbol of romance is essentially a pear-shaped diamond with a cleft at the top. The skill of the cutter determines the beauty of the cut. Look for a stone with an even shape and a well-defined outline.

Trilliant

Trilliant-cut gemstones are triangular shaped stones usually based on the brilliant cut, hence the name. The basic trilliant design has 43 facets, but modern variations may have 50 or more facets. Because of their equilateral form, trilliants return substantial light and color to the eye. Trilliants are considered nearly as brilliant as round cuts, so are a good choice for buyers who like brilliance but want something other than round.

Cabochon

This is a modification to the gemstone face that creates a highly polished, convex-cut, unfaceted gem that is smooth to the touch.

Part 7　Text

Gemstone Cut

Most of the gemstones are worn as waterworn and damaged or rounded pebbles, though some good crystals may also be obtained. The value and beauty of gemstones are very much enhanced by the proper cutting of facets, because the optical properties are then brought
out to the best advantage. In cutting, it is always the aim to maintain the symmetry of the crystal.

The quality of a gemstone's cut can have a dramatic impact on how it looks but only a small impact on the price per carat. Jewelers and savvy gem shoppers are paying more attention to cut to get the most beauty for the money.

How can you tell if a gem is well cut? The easiest way to train your eye is to look at bad cutting next to good cutting. Look at the picture above, which is of three rubies, all with good color and clarity. The ruby in the middle looks better because all the light that goes in is reflected back to the eye. See how the color is evenly distributed across the surface of the stone. This ruby also has more life and sparkle as the light dances across the facets. In contrast, the other two stones have dark areas where light is not reflected to the eye of the observer at the right angle.

Cutting style can also add beauty to a gemstone. Look at these two stones: they are the same size, shape, fine quality and color but dramatically different in appearance because one is a standard emerald cut and the other is a Barion cut, with more facets in the back.

In addition to these standard gemstone shapes, gem designers are inventing new ways of cutting gemstones in unique individual styles. For example, some facet gems in unusual geometrical shapes, some carve gemstones and some use a combination of faceting and carving.

A good cut showcases the gemstone's color, diminishes its inclusions, and exhibits good overall symmetry and proportion. Because gemstone color can vary, there are no hard geometrical standards when it comes to maximizing brilliance or color. Gemstones, especially

rarer ones, are sometimes cut for size without regard for their color. For example, when corundum varieties such as sapphire and ruby are cut for maximum weight rather than beauty, they may display banded colors or streaks.

In a gemstone with more saturated color, the best cut may be more shallow than average, permitting more light to penetrate the gemstone, while in a less saturated gem, the color may benefit from a deeper cut.

Look at the gemstone in the setting and ensure that all the facets are symmetrical. An asymmetrically cut crown indicates a gemstone of low-quality. In all cases, a well-cut gemstone is symmetrical and reflects light evenly across the surface, and the polish is smooth, without any nicks or scratches. Like diamonds, fine quality color gems usually have a table, crown, girdle, pavilion, and culet. Iridescent opals are one exception, and most often have a rounded cabochon cut.

Most gems that are opaque rather than transparent are cut as cabochons rather than faceted, including opal, turquoise, onyx, moonstone. You will also see lower grade material in gemstones such as sapphire, ruby and garnet cut as cabs. If the gem material has very good color but is not sufficiently transparent or clean to be faceted, it can still be shaped and polished to be a very attractive cabochon. It is also common to cut softer stones as cabs, since gems with a hardness less than 7 (on the Mohs' scale) can easily be scratched by the quartz in dust and grit. Minute scratches show much less on a cabochon than on a faceted stone.

Part 8　Words and Expressions

glare[glɛə]	n. 炫目的光
valentine['væləntaɪn]	n. 情人,情人节礼物
hesitate['hezɪteɪt]	v. 犹豫,踌躇,不愿
culet['kju:lɪt]	n. 底光
bead[bi:d]	n. 珠子
princess[prɪn'ses]	n. 公主
cabochon[kæbəʃɒn]	n. 弧面宝石
rough[rʌf]	adj. 粗糙的,粗略的,大致的,粗野的,粗暴的
	v. 大体描述　adv. 粗糙地
carve[kɑ:v]	v. 雕刻,切开
distinctive[dɪs'tɪŋktɪv]	adj. 与众不同的,有特色的
bond[bɔnd]	n. 结合,约定
calibrate['kælɪbreɪt]	v. 校准

Unit 9　Gemstone Cut

rectangular[rek'tæŋjulə]	adj. 矩形的
solitaire[ˌsɒlɪ'teə]	n. 单粒宝石,镶单粒宝石的戒指
flatter['flætə]	vt. 过分夸赞,奉承,阿谀,使高兴,使满意
concentric[kən'sentrɪk]	adj. 同中心的
broad[brɔːd]	adj. 宽的,阔的
symmetrical[sɪ'metrɪkəl]	adj. 对称的
elongate[iːlɒŋgeɪt]	v. 拉长,(使)伸长,延长
hybrid['haɪbrɪd]	n. 混合物,杂种,混血儿　adj. 混合的,杂种的
marquise[mɑː'kiːz]	n. 侯爵夫人,女侯爵
complement['kɒmplɪmənt]	n. 补足物,[义法]补语,[数]余角　vt. 补助,补足
cushion['kʊʃn]	n. 垫子
cleft[kleft]	n. 裂缝,隙口
equilateral[iːkwɪ'lætərəl]	adj. 等边的　n. 等边形
substantial[səb'stænʃəl]	adj. 坚固的,实质的,真实的,充实的
waterworn['wɔːtəwɔːn]	adj. 被水冲蚀的
pebble['pebl]	n. 小圆石,小鹅卵石
savvy['sævɪ]	v. 知道,了解　n. 机智,头脑,理解,悟性
Barion cut	重子切割（切割拥有传统阶梯切割冠部及改良明亮式切割亭部,除亭尖外,方形 Barion 切割的钻石共有 62 切面）
showcase['ʃəʊkeɪs]	n. (商店或博物馆的玻璃)陈列橱,显示优点的东西 vt. 使显出优点
proportion[prə'pɔːʃn]	n. 比例
streak[striːk]	v. 飞跑,加上条纹
asymmetrically[eɪsɪ'metrɪklɪ]	adv. 不对称
nick[nɪk]	n. 刻痕,缺口　vt. 刻痕于,挑毛病　vi. 阻击
grit[grɪt]	n. 粗砂

Part 9　Check Your Understanding

Exercise 1　Answer the following questions

(1) Which three segments is a cut-gemstone commonly parted into?
(2) What difference is between facet from cabochon in gem-cut-style?
(3) What is the round-brilliant-cut? What is the princess-cut?
(4) When are the gemstones cut into cabochon?
(5) How can you tell if a gem is well cut?

Exercise 2　Fill in the blanks

(1) Please illustrate several gem-cut styles: _____, _____, and _____.
(2) More than 75% of diamonds sold today are cut into _____ style.
(3) Jewelers and savvy gem shoppers are paying more attention to gem-cut because _____

_____.

(4) Gemstones, especially rarer ones, are sometimes cut for _____ without regard for their color.

Exercise 3 Translation

(1) Unlike diamonds, colored gemstones are not cut to achieve outstanding brilliance. This text offers advice to help you determine if a gemstone's cut exposes its beauty in the best possible way.

(2) When the final gem is made of two different materials it is called a "doublet", and when made with three, a "triplet". Common doublet or triplet gemstones are emerald and opal, usually combined with quartz or glass.

(3) A good cut showcases the gemstone's color, diminishes its inclusions, and exhibits good overall symmetry and proportion.

(4) 祖母绿型切工能很好地彰显宝石的颜色。

(5) 不透明的宝石如玛瑙、绿松石等一般都切磨成弧面型。

(6) 切磨好坏直接影响宝石的美丽程度和商业价值。

Part 10 Self-Study Material

Judging the Quality of the Setting

Now that you have found the perfect gem, all you need to do is make sure it is displayed well and held securely in place.

To judge the quality of the jewelry setting, pay close attention to details. Is the metal holding the stone even and smoothly finished so it won't catch on clothing? Is the stone held firmly and square in the setting? Is the metal well polished with no little burrs of metal or pockmarks?

Inexpensive jewelry often is very lightweight to give you a bigger look for the money. If a piece is lightweight, pay special attention to the prongs holding the stone: are they sturdy? Do they grip the stone tightly? You won't be happy about the money you saved in gold cost if you lose your stone!

If the piece is gold, does it have a karatage mark? Is the company trademark stamped next to it? If it is, the company is standing behind that mark and assuring you that the karatage is as stated.

When buying a necklace, make sure it lays well around the neck. Try it on or ask a salesclerk to model it so you can check how it fits against the skin. For earrings, check to make sure that they hang well from the ear and don't tip forward. Designs that are asymmetrical should have a left and right which mirror each other.

Here is one final hint from real jewelry buying pros: if a piece of jewelry is really well made, the back will be well finished also.

If you are buying a gift and you are not sure about the style of the piece of jewelry, why not give the perfect gemstone in a black velvet pouch and let the lucky recipient design her own perfect setting; many jewelers offer custom design services. Gems can speak louder than words. You can choose a gemstone that symbolizes what you want to say with your gift.

Unit 10　Diamond

Part 1　Dialogue

本章音频

Mary：Good afternoon, Madam. May I help you?
Kate：I'd like to buy a diamond ring for myself.
Mary：Are you interested in any special brand?
Kate：No. Match is the most important, I think.
Mary：There are many styles of diamond ring and the price varies from hundreds of dollars to thousands of dollars.
Kate：I don't know how you make the diamond's price.
Mary：Oh, You can think over the 4C's, color, clarity, cut and carat weight.
Kate：Would you mind speaking more slowly?
Mary：Yes. You may look at these identification certificates; the grade of diamond is reported on.
Kate：I'm confused. You'd better find a diamond fitting me well. I believe you.
Mary：Platinum or K-gold?
Kate：I prefer noble platinum.
Mary：Well. This is H-color, VS-clarity, excellent-round-brilliant cut, 1.02ct platinum diamond ring. Would you try this one?
Kate：Oh, how nice! Do you think it'll look good on me?

Mary: Certainly, Madam. It appeared to be very exquisite and artistic.
Kate: Thanks a lot. I'll take it.

Part 2 Fill the Blanks Basing on Your Gemological Knowledge

Gemological characteristics of diamond:

chemical composition: ____　　crystal system: ____　　Mohs' hardness: ____

specific gravity: _____　　color: _____

refractive index: _____　　dispersion: _____

Part 3 Link the Words and the Relevant Pictures

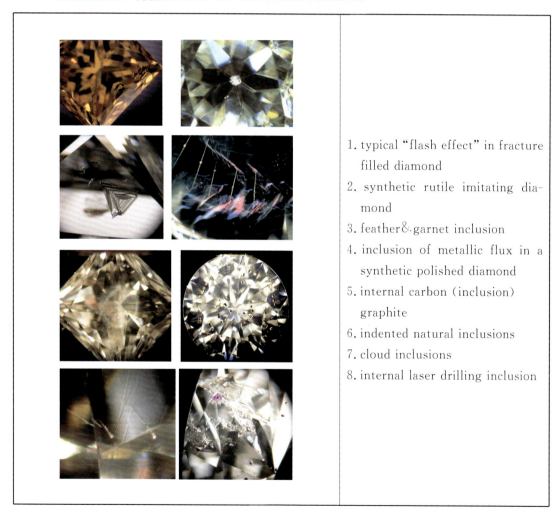

1. typical "flash effect" in fracture filled diamond
2. synthetic rutile imitating diamond
3. feather & garnet inclusion
4. inclusion of metallic flux in a synthetic polished diamond
5. internal carbon (inclusion) graphite
6. indented natural inclusions
7. cloud inclusions
8. internal laser drilling inclusion

Part 4 Useful Phrases

diamond grading　　钻石分级

diamond master set　　比色石

masterstone of fluorescence degree　　荧光强度对比样品

internal (external)characteristics　钻石的内部(外部)特征
round brilliant cut　标准圆钻型切工
table facet　台面
upper main facet　冠部主刻面(风筝面)
finish of diamond cut　钻石切工修饰度
loose diamond　裸钻,已抛磨但未镶嵌的钻石
fancy diamond necklace　花式钻石项链
engagement diamond ring　订婚钻戒
A diamond is forever(Debeers)　钻石恒久远,一颗永流传(戴比尔斯)

Part 5　Some Sentences

(1) To make it simple, the larger the diamond, the rarer it is (carat weight); the purer the diamond, the more valuable it is (clarity); the less color in a diamond, the more beautiful; and the more precise the cut of the diamond, the more brilliant it is.

(2) A carat is the universal measure of weight for a diamond. It's the easiest of the 4C's to determine, but two diamonds of equal size can have very different values, because the quality is still determined by the color, clarity and cut.

(3) "A Diamond Is Forever" is internationally known to men and women of all ages and may very well mean something different for everyone.

(4) Clarity is an indication of a diamond's purity. In all diamonds, except the most rare, tiny traces of minerals, gases, or other elements were trapped inside during the crystallization process. These are called inclusions, but are more like birthmarks. They may look like tiny crystals, clouds, or feathers and they're what make each diamond different and unique.

Part 6　Short Paragraph

Diamond Grading and the 4C's

A diamond's cost is based on the characteristics known as the "4C's". Clarity, Color and Cut (proportion) are the quality elements which together with the Carat Weight determine the value of a stone. Remember, each "C" is important in contributing to a diamond's quality. For example, a colorless diamond is at the top color, but if it lacks clarity, is small, or is not well-cut, it will be of lower value. The finest diamonds possess the rarest quality in each of the 4C's, and are the most valuable.

Clarity

A diamond's clarity is determined by the number, nature, position, size and color of internal characteristics called "inclusions" and surface features called "blemishes". All diamonds have identifying characteristics, but most are invisible to the naked eye. To view a diamond, experts use a $10\times$ magnifying loupe which allows them to see the appearance of tiny crystals, feathers or clouds. There are five categories in class that the final clarity grade is usually determined by how easy the inclusions and blemishes are for the grader to see.

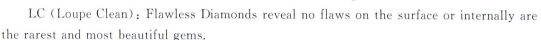

LC (Loupe Clean): Flawless Diamonds reveal no flaws on the surface or internally are the rarest and most beautiful gems.

VVS(Very Very Slightly Included): Contain minute inclusions that are difficult for even a skilled grader to locate under $10\times$. VVS_1: extremely difficult to see, visible only from the pavilion or small and shallow enough to be removed by minor re-polishing. VVS_2: very difficult to see.

VS (Very Slightly Included): Contain minor inclusions ranging from difficult (VS_1) to somewhat easy (VS_2) for a trained grader to see under $10\times$.

SI (Slightly Included): Contain noticeable inclusions which are easy (SI_1) or very easy (SI_2) to see under $10\times$. In some SIs, inclusions can be seen with the unaided eye.

P (Included): Contain inclusions which are obvious to a trained grader under $10\times$, can often be easily seen face-up with the unaided eye, seriously affect the stone's potential durability, or are so numerous they affect transparency and brilliance.

Color

Ideally, a diamond should have no color at all, like a drop of spring water. Increasing degrees of body color are measured on a scale ranging from no color at all (D) to deeply colored (Z). Beyond "Z" is the range where the diamond's color is vivid and rich, called "fancy colors". Diamonds of known color are used as comparison stones for color grading. Grading is done by comparing the diamond to be graded against these "master stones" under either artificial or natural north daylight (in the Northern Hemisphere). A machine called the "Colorimeter" can be used for color grading but there is no substitute for the trained human eye.

Carat Weight

Carat is the unit of weight for all gemstones. One carat is subdivided into 100 "points". Therefore a diamond measuring 75 points is 3/4 carat in weight, or 0.75ct. There are five carats in a gram. The word "carat" comes from the seed of the carob tree pod which is found in tropical climates. These seeds were used until this century to weigh precious gems.

Cut

When gemologists say "cut", they are talking about a gemstone's proportions, such as its depth and width and the uniformity of its facets—all characteristics that control brilliance, durability and other features we look for in a diamond.

Cut ensures that a given stone has maximum brilliance and sparkle which would not be the case were the stone cut for weight alone. Cut actually refers to two aspects of a diamond. The first is its shape (round, marquise, etc), the second is how well the cutting has been executed.

Part 7　Text

Gemological Characteristics of Diamond

Commercially, diamond is the most important of all gem species. It is estimated that diamonds account for approximately 90% of the value of gemstones purchased throughout the world. Diamond is always faceted to display its unique combination of adamantine luster and fire. Its supreme hardness ensures a lasting precision of cut which is unique among gemstones.

A common classification is industry diamond and jewelry diamond. The latter is usually graded by the "4C" criteria: carat, clarity, color and cut. In addition to the regular round, diamonds are often cut like pear, oval, heart, horse-eye, triangle, rectangle, or emerald square. Elaborately designed, diamonds can be set to make all types of precious jewels such as necklace, earring, ring and etc. Gemological characteristics of diamond are listed below:

(1) Chemical composition: C (carbon).

(2) Crystal system: Cubic.

(3) Habit: Most important is the octahedron. Diamond also occurs as cubes, dodecahedron, modified cubes. Crystals are frequently distorted, and crystal faces may be curved. Twinned octahedral crystals (macles are common).

(4) Surface features: Triangular etch figure may be seen on octahedral faces. Cleavage: Perfect octahedral. Cleavage may be used in the fashioning of diamond, to split large crystals or trim off flawed material. Seen in and on cut and rough stones.

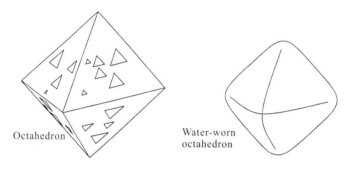

Octahedron   Water-worn octahedron

(5) Hardness: 10. Diamond is the hardest known natural substance. The hardness of diamond varies according to the crystallographic orientation. But in any direction, diamond is still much harder than any other gemstone.

(6) Specific gravity: 3.52.

(7) Color: Colorless; yellowish, brownish or greenish. Fancy colors (those of a distinct hue) include yellow and brown, rarely green, pink and blue, very rarely red and purple. Except for stones of fancy color, the desirability and value of diamond decreases as the depth of hue increases.

(8) Luster: Adamantine.

(9) Refractive index: 2.42 single. Many diamonds display anomalous extinction.

(10) Dispersion: High 0.044. Diamond displays a higher degree of dispersion than any other natural colorless gemstone.

(11) Luminescence: The fluorescence of diamond varies in color and intensity. It is stronger in long wave than short wave ultraviolet light. The color displayed may be bluish-white to violet, greenish or yellowish. Some stones are almost inert. The variability of dia-

mond fluorescence is particularly useful in appraising jewelry set with many colorless stones. If all stones set in a piece show similar fluorescence the stones are unlikely to be diamond.

Those diamonds which fluoresce blue under ultra violet light may show yellow phosphorescence. This is diagnostic for diamond.

(12) Occurrence: Mainly from description of kimberlite pipes or alluvial deposits.

(13) Localities: The alluvial deposits of India were the only known source of diamond from classical times until the eighteenth century. The important Brazilian fields were discovered in about 1725.

The alluvial and kimberlite pipe deposits of South Africa were discovered in the latter half of the nineteenth century, and those in Siberia during the 1940s. More recently Australia has become an important producer of diamonds which are found in lamproite, a rock similar to kimberlite.

Important gem diamond producing countries include Angola, Australia, Botswana, Brazil, China, Namibia, Russia, Sierra Leone, South Africa, Tanzania.

(14) Imitations: Many natural, synthetic and artificial products have been used to imitate diamonds. Of these, the most convincing, in terms of appearance, are:

① cubic zirconia (CZ).

② yttrium aluminate known as yttrium aluminium garnet or YAG.

③ colourless zircon.

④ some types of glass.

CZ is by far the best and most widely used simulant today. Other gemstones used as imitations include natural and synthetic white sapphire and synthetic white spinel.

Part 8 Words and Expressions

brand[brænd]	n. 商标,牌子,烙印 vt. 打火印
confuse[kənˈfjuːz]	vt. 搞乱,使糊涂
exquisite[ˈekskwɪzɪt]	adj. 优美的,高雅的,精致的,剧烈的,异常的,细腻的,敏锐的
chemical composition	化学成分
crystal system	晶系
Mohs' hardness	摩氏硬度
flash effect	闪光效应
fracture filled diamond	裂隙充填钻石

metallic flux	金属熔剂
graphite['græfaɪt]	n. 石墨
indented[ɪn'dentɪd]	adj. 锯齿状的,犬牙交错的
cloud inclusion	云状包裹体
laser drilling	激光钻孔
precise[prɪ'saɪs]	adj. 精确的,准确的　n. 精确
brilliant['brɪljənt]	adj. 灿烂的,闪耀的,有才气的
universal[ˌjuːnɪ'vɜːsəl]	adj. 普遍的,全体的,通用的,宇宙的,世界的
internationally[ˌɪntə'næʃənlɪ]	adv. 国际性地,在国际间
indication[ˌɪndɪ'keɪʃn]	n. 指出,指示,迹象,暗示
crystallization['krɪstəlaɪ'zeɪʃn]	n. 结晶化
birthmark['bɜːθmɑːk]	n. 胎记,胎痣
proportion[prə'pɔːʃn]	n. 比例,均衡,面积,部分
	vt. 使成比例,使均衡,分摊
contribute[kən'trɪbjuːt]	v. 捐助,捐献,贡献,投稿
lack[læk]	n. 缺乏,短缺的东西　vt. 缺乏,没有,需要
	vi. 缺乏,没有
naked eye	肉眼
10× magnifying loupe	10倍放大镜
LC (Loupe Clean)	镜下无瑕
flaw[flɔː]	n. 缺点,裂纹
VVS(Very Very Slightly Included)	非常非常小的瑕疵
extremely[ɪks'triːmlɪ]	adv. 极端地,非常地
VS(Very Slightly Included)	非常小的瑕疵
a trained grader	训练有素的分级师
SI(Slightly Included)	小瑕疵
P(Included)	重瑕疵
potential[pə'tenʃl]	adj. 潜在的
beyond[bɪ'jɔnd]	prep. 超过
fancy colours	彩色
master stones	（钻石）比色石
colorimeter[ˌkʌlə'rɪmətə]	n. 色度计,色量计
gram[græm]	n. 克
tropical['trɔpɪkl]	adj. 热带的
climate['klaɪmɪt]	n. 气候
sparkle['spɑːkl]	v. 发火花
commercially[kə'mɜːʃəlɪ]	adv. 商业上
approximately[ə'prɔksɪmətlɪ]	adv. 近似地,大约
adamantine lustre	金刚光泽

supreme[sə'pri:m]   adj. 极度的,极大的,至高的,最高的
triangle['traɪæŋgl]   n. 三角形
rectangle['rektæŋgl]   n. 长方形,矩形
elaborately[ɪ'læbərətlɪ]   adv. 苦心经营地,精巧地
octahedron[ˌɔktə'hɪdrən]   n. 八面体
cube[kju:b]   n. 立方体
dodecahedron[ˌdəudɪkə'hedrən]   n. 十二面体
distorted[dɪs'tɔ:tɪd]   adj. 扭歪的
twinned octahedral crystals   八面体双晶
macle['mækl]   n. 双晶
triangular[traɪ'æŋgjulə]   adj. 三角形的,三人间的
etch figure   蚀象
fashioning['fæʃənɪŋ]   n. 精加工
fashioning of diamond   钻石的加工
trim off   修剪
crystallographic orientation[krɪstələ'græfɪk] [ɔ:rɪən'teɪʃn]   结晶方向
anomalous[ə'nɔmələs]   adj. 不规则的,反常的
extinction[ɪks'tɪŋkʃn]   n. 消光
luminescence[ˌlu:mɪ'nesns]   n. 发光
ultraviolet light   紫外光
inert[ɪ'nə:t]   adj. 惰性的
pipe[paɪp]   n. 管,导管,烟斗,管乐器,笛声
                                             vt. 以管输送,吹哨子 vi. 吹笛,尖叫
kimberlite pipe   金伯利岩岩管
alluvial[ə'lu:vɪəl]   adj. 冲积的
deposit[dɪ'pɔzɪt]   n. 堆积物
alluvial deposits   冲积物,冲积矿床
Siberia[saɪ'bɪərɪə]   n. 西伯利亚
lamproite['læmprɔɪt]   n. 钾镁煌斑岩
Angola[æŋ'ɡəulə]   n. 安哥拉
Sierra Leone['sɪərə] [li:'əun]   n. 塞拉利昂(非洲)
imitation[ˌɪmɪ'teɪʃn]   n. 赝品,仿造物
YAG   人造钇铝榴石

Part 9 Check Your Understanding

Exercise 1   Answer the following questions

(1) Please tell gemological characteristics of diamond.
(2) Which countries was the diamond produced?
(3) What is the meaning of "VVS" in diamond grade?

(4) What is the meaning of "D" in diamond grade?

(5) What is the "carat"?

Exercise 2　Fill in the blanks

(1) The jewelry diamond is usually graded by the four "C" criteria. The 4C is _____ _____.

(2) _____ is the universal measure of weight for a diamond.

(3) Diamond clarity grade is classed into five categories: _____.

(4) _____ is the best and most widely used diamond imitation today in market.

Exercise 3　Translation

(1) Commercially, diamond is the most important of all gem species. It is estimated that diamonds account for approximately 90% of the value of gemstones purchased throughout the world.

(2) A diamond's cost is based on the characteristics known as the "4C's". Clarity, Color and Cut (proportion) are the quality elements which together with the Carat Weight determine the value of a stone.

(3) Carat is the unit of weight for all gemstones. One carat is subdivided into 100 "points". There are five carats in a gram. The word "carat" comes from the seed of the carob tree pod which is found in tropical climates. These seeds were used until this century to weigh precious gems.

(4) 钻石的价值取决于4C,即颜色、净度、克拉质量和切工。

(5) 彩色钻石和异形切工钻石不在一般的 4C 分级之列。

(6) 钻石因色散高而出火、硬度高而耐久、产出少而名贵被誉为"宝石之王"。

Part 10　Self-Study Material

How to Clean Diamond Jewelry

　　Hand lotions, hair styling products and everyday grime all leave enough of a film on your diamond ring to keep it from looking its best. And if you wait too long between cleanings, those materials can accumulate into a thick layer of gunk on the back of your diamond, blocking light and making the diamond appear dull and lifeless.

　　Diamonds are the hardest substance known, but that doesn't mean we can bring them back to life with any old cleanser. Coatings and other materials used to enhance diamonds can sometimes be removed by harsh chemicals or vigorous scrubbing, so take care when it's time to make your diamond ring sparkle.

Gentle & Effective Ways to Clean Diamond Rings

　　(1) Soak your diamond ring in a warm solution of mild liquid detergent and water. Ivory dishwashing liquid is a good choice, but any other mild detergent is fine.

　　(2) Use a soft brush if necessary to remove dirt. Soft is the key—don't use a brush with

bristles that are stiff enough to scratch the ring's metal setting.

(3) Swish the ring around in the solution, and then rinse it thoroughly in warm water. Close the drain first, or put the ring in a strainer to keep from losing it!

(4) Dry the diamond ring with a lint-free cloth.

Cleaning Unfilled Diamonds

Diamonds that have not been fracture-filled can be cleaned with a solution of ammonia and water.

Use the gentler liquid detergent solution for fracture filled diamonds, because ammonia might eventually either cloud or remove the coating that's been placed on the gemstone.

Cleaning Rings with Multiple Types of Gemstones

The method you use to clean jewelry should protect its weakest element. If your ring includes other gems, use a cleaning method that is suitable for the less durable stones.

Protect Diamond Rings from Chlorine

You might already protect your hands from harsh chemicals, but if you don't, think about how chemicals such as chlorine can affect your fine jewelry. Remove your rings or wear gloves to keep chlorine away from your rings.

Unit 11   Ruby and Sapphire

a 2.5 kg parcel of rough sapphires   一包 2.5kg 的蓝宝石原石
rutile silk in a Vietnamese ruby   越南红宝石中的金红石丝状包
    裹体
a heat-treated Mong Hsu ruby   缅甸孟宿热处理红宝石
pigeon's blood ruby   鸽血红红宝石
Burma Mogok Mine   缅甸抹谷矿
aluminium oxide (corundum)   氧化铝 (刚玉)
hexagonal crystal system   六方晶系
metamorphic rocks, dolomitic marble and gneiss   变质岩、白云大理岩和片麻岩

Part 5   Some Sentences

(1) Oriented rutile crystal inclusions cause a six-rayed-star light effect (called asterism) to form the popular star ruby.

(2) Sapphires are well known among the general public as being blue, but can be nearly any color.

(3) Rutile, with an acicular habit, is a common inclusion in ruby or sapphire that creates a chatoyancy, cat's eye effect, or asterism when the gem is oriented and cut properly.

(4) Thailand is the world's most important ruby trading center. Perhaps 80 percent of the world's ruby goes through Thailand at some point in the trading cycle. The largest ruby cutting factories are in the Chanthaburi area of Thailand. Bangkok is generally where the world's buyers come to purchase ruby.

Part 6   Short Paragraph

The Color of Ruby Gemstones

Which color would you spontaneously associate with love and vivacity, passion and power? It's obvious, isn't it? Red. Red is the color of love. It radiates warmth and a strong sense of vitality. And red is also the color of the ruby, the king of the gemstones. The ruby is the perfect way to express powerful feelings. Jewelry with a precious rubies attests to that passionate,  intense love that people can feel for each other. The brightest and best red in ruby colors is called Pigeon Blood Red. True pigeon's blood red is extremely rare, produced in Burma Mogok Mine, more a color of the mind than the material world. One Burmese trader ex-

pressed it best when he said: "Asking to see the pigeon's blood is like asking to see the face of God."

For thousands of years, the ruby has been considered one of the most valuable gemstones on earth. It has everything a precious stone should have: magnificent color, excellent hardness and outstanding brilliance. In addition to that, it is an extremely rare gemstone, especially in its finer qualities.

Color is a ruby's most important feature. Its transparency is only of secondary importance. So inclusions do not impair the quality of a ruby unless they decrease the transparency of the stone or are located right in the centre of its table; on the contrary, inclusions within a ruby could be said to be its "fingerprint", a statement of its individuality and, at the same time, proof of its genuineness and natural origin. The cut is essential: only a perfect cut will underline the beauty of this valuable and precious stone in a way befitting the "king of the gemstones". However, a really perfect ruby is as rare as perfect love. If you do come across it, it will cost a small fortune. But when you have found "your" ruby, don't hesitate: hang on to it!

Part 7 Text

Gemological Characteristics of Ruby and Sapphire

Ruby and sapphire are the gem quality of the mineral corundum. They are both rated among the five most costly stones. Ruby is the July birthstone and 15th and 40 anniversary stone. Sapphire is the September birthstone and for the 5th or 45th anniversary gemstone gifts. Gemological characteristics of ruby and sapphire are listed below:

(1) Chemical composition: Al_2O_3 aluminium oxide.

(2) Crystal system: Trigonal.

(3) Habit & surface features: Prismatic or tabular six-sided crystals, with flat basal terminations. Rhombohedral faces may be developed at alternate corners. Also occurs as six-sided bipyramids, with varying degrees of pyramid angle giving long thin and short thick bipyramids. Sometimes, corundum may show itself a barrel shaped appearance. Pyramids and prisms may show striations at 90° to the c-axis. Triangular growth markings are fre-

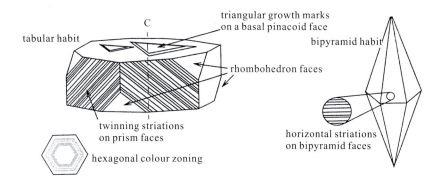

quently seen on pinacoid faces. These provide good recognition features.

Corundum commonly occurs as water-worn pebbles; these may show signs of twin planes on the surface.

(4) Cleavage: No cleavage, but conchoidal or splintery fracture. "Parting" of twin planes occurs parallel to the basal and rhombohedral faces.

(5) Hardness: 9.

(6) Specific Gravity: 4.

(7) Luster: Vitreous to sub-adamantine.

(8) Refractive index: 1.76 to 1.78.

(9) Birefringence: 0.008.

(10) Optical nature: Uniaxial.

(11) Color and Variety: In a pure state corundum is colorless, but can also be red, blue, pink, orange, yellow, green, purple and black. Gem quality red corundum is known as ruby, whereas any other color is termed sapphire.

(12) Pleochroism: Ruby—Strong-deep red and orange-red. Sapphire—Strong in varieties other than yellow. Other colors will show differing shades of the body color.

(13) Spectrum: Ruby—The color is caused by the presence of chromium, and the full diagnostic spectrum consists of a double line plus two weaker lines in the red, general absorption of the yellow and green, three fine lines in the blue, and general absorption of the violet. With a hand spectroscope only two of the lines in the blue are normally seen. The double line and occasionally the other lines in the red, will be seen as bright emission lines, instead of absorption lines; Sapphire—The absorption spectrum of blue sapphire usually consists of a group of three bands in file blue. A similar spectrum is seen in many green and golden stones.

(14) Inclusions: Solid crystal inclusions found within corundum include apatite, calcite, chlorite, corundum, fluorite, graphite, olivine, pyrite, quartz, spinel, tourmaline, zircon and so on. Other inclusions include liquid/gas/solid filled cavities, that create "wispy" pat-

terns, which may be confused with flux inclusions of synthetics; growth zoning, which are straight lines that follow the angular, hexagonal pattern reflecting the crystal system and form; and exsolved solid inclusions such as rutile needles and hematite-ilmenite plates. The exsolved rutile inclusions are called "silk" and are oriented in three directions at between 60 and 120 degree in the basal plane.

(15) Localities: Ruby is found in commercial quantities in many locations, the most important of which are: Myanmar, Vietnam, Pakistan, Afghanistan, Thailand, Cambodia, Sri Lanka, Tanzania. Sapphire major areas include Sri Lanka, Australia, Kashmir, Myanmar, Thailand, Cambodia, USA.

(16) Luminescence: Rubies, and pink and purple sapphires all contain chromium and, therefore, fluoresce red in ultraviolet light. Natural sapphires often contain traces of iron, and this may sometimes mask their fluorescence.

(17) Star rubies and sapphires: natural star stones with good body color and sharp stars are rare. Generally a six rayed star is seen, rarely 12 rays, when rutile needle inclusions are present in sufficient quantities, and orientated in directions parallel to the lateral crystal axes. Stones must be cut as cabochons with the plane of the inclusions parallel to the base of the cabochon. Natural star stones usually contain much coarser rutile needles than synthetic star corundum.

(18) Synthetic Corundum: Synthetic corundum is made in all colors, by various methods, each of which produces characteristic inclusions.

(19) Artificial Treatment: Most ruby and sapphire is heat-treated to improve the color. Sapphires, and to a lesser extent rubies, may be surface diffusion treated to alter or enhance the color to one which is more desirable. Rubies may be treated with red oil which enters fractures and improves the color and clarity. Fractures, cavities and fissures in ruby may be filled with glass.

Part 8 Words and Expressions

fancy[ˈfænsɪ]	adj. 奇特的,异样的 vt. 想象,设想,爱好
	n. 爱好,迷恋,想象力
bangle[ˈbæŋgl]	n. 手镯
luxuriant[lʌgˈʒuːrɪənt]	adj. 奢华的
artistic[ɑːˈtɪstɪk]	adj. 艺术的,有美感的,风雅的
Ceylon[sɪˈlɒn]	n. 锡兰(印度以南一岛国,现已更名为斯里兰卡 Sri Lanka)
Ceylon sapphire	斯里兰卡(锡兰)蓝宝石

Unit 11　Ruby and Sapphire

cabochon sapphire	弧面型蓝宝石
boule[buːl]	n. 梨形人造宝石
synthetic ruby boules	合成红宝石梨晶
curve[kəːv]	n. 曲线，弯曲
flame fusion	焰熔法（合成宝石）
oriented rutile	定向排列的金红石
six-rayed-star light effect (called asterism)	六射星光效应（星光效应）
acicular[əˈsɪkjulə]	adj. 针状的，针尖状的
cat's eye effect	猫眼效应
Thailand [ˈtaɪlænd]	n. 泰国
Bangkok[bæŋkɔk]	n. 曼谷（泰国首都）
spontaneously[spɔnˈteɪnɪəslɪ]	adv. 自然地，本能地
vivacity[vɪˈvæsɪtɪ]	n. 活泼
attest[əˈtest]	v. 证明
passionate[ˈpæʃənɪt]	adj. 充满热情的
magnificent[mægˈnɪfɪsnt]	adj. 华丽的，高尚的，宏伟的
impair[ɪmˈpɛə]	v. 削弱
genuineness[ˈdʒenjuɪnnɪs]	n. 真实，真正，诚恳
anniversary[ˌænɪˈvɜːsərɪ]	n. 周年纪念
aluminium oxide	氧化铝
trigonal[ˈtrɪgənəl]	adj. 三角形的
prismatic[prɪzˈmætɪk]	adj. 棱镜的
tabular[ˈtæbjulə]	adj. 制成表的，扁平的，表格式的，平坦的
	vi. 列表，排成表格式
rhombohedral[ˌrɔmbəuˈhiːdrəl]	adj. 菱面体的
six-sided bipyramids	六方双锥
pinacoid[ˈpɪnəkɔɪd]	n. 轴面
water-worn	水蚀的
pebble[ˈpebl]	n. 小圆石，小鹅卵石
conchoidal[kɔŋˈkɔɪdl]	adj. 贝壳状的
splintery[ˈsplɪntərɪ]	adj. 裂片似的，易碎裂的，碎裂的
vitreous to sub-adamantine	玻璃—亚金刚的
uniaxial[ˌjuːnɪˈæksɪəl]	adj. 单轴的
hand spectroscope	手持分光镜
emission[ɪˈmɪʃn]	n. （光、热等的）散发，发射，喷射
emission lines	（分光镜上的）吸收暗线
chlorite[ˈklɔːraɪt]	n. 绿泥石
pyrite[ˈpaɪraɪt]	n. 黄铁矿
wispy[ˈwɪspɪ]	adj. 像小束状的，纤细的，脆弱的

flux inclusions	熔剂包裹体
ilmenite[ˈɪlmɪnaɪt]	n. 钛铁矿
Myanmar[ˈmiːənmɑː]	n. 缅甸(东南亚国家)(即 Burma)
Vietnam[vjetˈnæm]	n. 越南
Pakistan[ˌpækisˈtæn]	n. 巴基斯坦(南亚国家)
Afghanistan[æfˈgænɪstæn]	n. 阿富汗(西南亚国家)
Cambodia[kæmˈbəudɪə]	n. 高棉,柬埔寨
Sri Lanka[sriˈlæŋkə]	n. 斯里兰卡(南亚岛国)
Kashmir[ˈkæʃmɪə]	n. 克什米尔(南亚一地区,约 2/5 为巴基斯坦控制,其余为印度控制)
traces of iron	微量的铁
coarse[kɔːs]	adj. 粗糙的,粗鄙的

Part 9　Check Your Understanding

Exercise 1　Answer the following questions

(1) Please tell gemological characteristics of ruby.
(2) Which countries was the sapphire produced?
(3) What is the "pigeon's-blood ruby"?
(4) Which mineral inclusions are commonly within ruby and sapphire?
(5) Which artificial treatments is for ruby and sapphire?

Exercise 2　Fill in the blanks

(1) The best ruby is produced in _____ .
(2) _____ is the universal measure of weight for a diamond.
(3) _____ and _____ are the variations gemstones of mineral corundum.
(4) The color of ruby is caused by _____ .

Exercise 3　Translation

(1) Color is a ruby's most important feature. Its transparency is only of secondary importance. So inclusions do not impair the quality of a ruby unless they decrease the transparency of the stone or are located right in the centre of its table.
(2) Ruby is the July birthstone and 15th and 40 anniversary stone. Sapphire is the September birthstone and for the 5th or 45th anniversary gemstone gifts.
(3) Natural rubies and purple sapphires often contain traces of iron, and this may sometimes mask their fluorescence.
(4) 把含有大量针状包裹体的蓝宝石切成弧面,可能产生六射星光。
(5) 红宝石是指红色调的宝石级刚玉,由微量铬致色。
(6) 珠宝市场上可见合成的及优化处理的红宝石和蓝宝石。

Part 10 Self-Study Material

Sapphire Introduction

Sapphire is lubricious, transparent gem grade corundum. Actually, the nature of gem grade corundum except the red ruby says the various colors such as blue, green, yellow, gray, colorless, etc., are called sapphire. Color with Indian "cornflower blue" is the best. The ancient Persians believed the earth rests on a giant sapphire. Its reflection, they said, made the sky blue.

Sapphire is the original "true blue": the gem of fidelity and of the soul. In ancient times, a gift of a sapphire was a pledge of trust, honesty, purity, and loyalty. This tradition makes sapphire a popular choice for engagement rings.

The biggest characteristic of sapphire is inhomogenous color piece of clustering. Burma bright blue sapphire contains inclusions, can produce six or twelve shooting stars shooting. Indian Kashmir cornflower blue sapphire, is the indigo blue, purple bright-colored color, high quality sapphire. Sapphires in Sri Lanka, Thailand, China and Australia are all distinctive.

If you want a sapphire whose color is natural, insist on a report from a recognized gem testing laboratory like the AGTA Gemological Testing Center indicating the stone shows no evidence of heat alteration. But also be prepared to pay a premium for the privilege.

It is also recommended that you ask for lab verification of origin for any sapphire purported to be from Kashmir, Burma, or any other prestigious country of origin. Stones proven to be from certain localities command a premium based on that fact. GIA lab also offers respected origin reports.

Sapphire occurs readily in sizes up to 2 carats, and yet, it is not unusual to see gemstones in sizes of 5 to 15 carats. Sapphires are most common in cushion and oval shapes. Other shapes, including rounds, emerald cuts, princess cuts are readily available in sizes under a carat.

Sapphire has a hardness of 9 on the Mohs' scale and is also extremely tough. In its common form, it is even used as an abrasive. As a result, sapphires are the most durable of gems. Clean with mild dish soap; use a toothbrush to scrub behind the stone where dust can collect.

Unit 12   Emerald

Part 1 Dialogue

Kate: Excuse me. Can I try this emerald necklace on?

Mary: Yes, sure. It suits you well, I think. Why not try on the earrings too? They are whole set jewelry.

Kate: The size of the earrings seems too small, isn't it?

Mary: They are high-grade emeralds from Colombia. The matching is not a problem.

Kate: Is it of the highest quality?

Mary: Sure. The craft is also unique.

Kate: Well, is there discount?

Mary: How much do you want for this?

Kate: Half of the price.

Mary: Oh, No. The highest discount is 10% on this set. Our brand is famous, you know.

Kate: The price is beyond my budget. I'll buy it right away if it were cheaper.

Mary: Let me count. 15% is off, ok?

Kate: Yes. The emerald is attractive. Can you give me the invoice?

Mary: Sure. Please pay it at the cashier's over there at the first.

本章音频

Kate: Thank you.

Part 2 Fill the Blanks Basing on Your Gemological Knowledge

Gemological characteristics of emerald:
chemical composition:_____   mineral:_____   Mohs' hardness:_____
specific gravity:_____   color is caused by:_____
refractive index:_____   brittle:_____

Part 3 Link the Words and the Relevant Pictures

1. the classic three-phase inclusion of a Colombian emerald
2. loose cabochon emerald
3. the classic nail inclusion of a synthetic emerald
4. emerald pendant
5. emerald eardrop
6. emerald crystal
7. emerald ring

Part 4 Useful Phrases

aquamarine cat's eye 海蓝宝石猫眼
natural emerald round bead bracelet 天然祖母绿手链
emerald fake or doublet 仿祖母绿或祖母绿拼合石
poor basal cleavage (seldom visible), conchoidal fracture 偶见一组不完全解理，贝壳状断口
chatham (flux-growth) synthetic emerald 查塔姆(助熔剂)合成祖母绿
an incredibly fine 5-carat emerald crystal from Muzo mining district of Colombia 一粒来自哥伦比亚木佐矿区的极品5克拉的祖母绿晶体
beryllium aluminium silicate 铍铝硅酸盐

loose emerald certificate   祖母绿裸石鉴定证书

Trapiche variety emerald   达碧兹祖母绿

a fine quality cabochon natural emerald from Chivor   一粒质量上好、来自契沃尔的弧面祖母绿

the classical emerald cut   经典的祖母绿琢型

a bluish-green emerald with fine quality finish   一粒超精加工的蓝绿色祖母绿

faceted from a genuine emerald crystal mined at Muzo, Colombia   由哥伦比亚木佐矿区天然祖母绿晶体切割的刻面祖母绿

colorless-oiling emerald   浸油无色祖母绿

the typical of chromium-colored absorption spectrum   典型的铬吸收谱

Part 5　Some Sentences

(1) The emerald cutting centers are mainly based in Thailand and India. The largest emerald purchasing countries, accounting for almost 75% of the total purchase in the world are US and Japan.

(2) A large size emerald is rare thus the price per carat of emerald will significantly go up with the increase in the size of the gem.

(3) In colored stones transparency and clarity are closely linked. This is especially true with emeralds. Emeralds are normally found to have visible inclusions. Thus it is an accepted practice in the market to sell emeralds with inclusions or flaws.

(4) Emeralds are one of the most difficult gems to cut. The cutter should be extremely careful and plan the cut before he can start any process. Emeralds by nature have a lot of fractures (or cracks) in them and this makes the stone very brittle.

Part 6　Short Paragraph

General Information of Natural Emerald Gemstone

The word "emerald" brings to the mind the lush green color of a paradisiacal landscape. For thousands of years people have loved and admired the deep green color of the natural

emerald gemstone, symbol of natural beauty, thus making it one of the most popular gems in spite of its brittle nature and difficulties found in setting the gem for emerald jewelry. Emerald Gemstone is also called as the May Birthstone.

　　This gemstone gets it name from a Persian word "Esmeralde". In the history of emeralds the earliest know area where the natural emerald was found was Egypt, by the Red sea. The Egyptian emerald mines are in the hillside of Djebel Sikeit and Djebel Zabarah. Later, since high quality gemstones where mined from the Colombian mines, the Egyptian mines lost their importance (most were forgotten for many centuries). Big and precious roughs were also used in the form of emerald tablet.

Top quality special emeralds come from Colombia. So far, Colombian emeralds fetch the most unbelievable prices in the global market. This gem is also found in India, Zambia, Pakistan, Afghanistan, Russia, South Africa, Egypt, Zimbabwe, Austria, Brazil, Australia, Tanzania and Madagascar.

In fact, it is said that when the Colombian emeralds were discovered, people preferred suffering torture or even dying than to reveal the source of the mines. Such was the beauty and quality of emerald gemstone that were found in Colombia.

In ancient times (about 4000 BC in Babylon, the oldest known gem market), this fine quality emerald crystal gem was dedicated to the Goddess Venus. This gem represents immortality and faith, and thus a symbol for lovers who wish to express their undying love and faith. That is one of the reasons why it is one of the preferred stones for an engagement ring. Even on a wedding, they form a good contrast with the white wedding gown.

Part 7 Text

Gemological Characteristics of Emerald

Emerald, the green variety of the mineral beryl, is the most famous and favored green gemstone. Its beautiful green color, combined with durability and rarity, make it one of the most valuable gemstones. In top quality, fine emeralds are even more valuable than diamonds.

The chemical formula of emerald gemstone is beryllium aluminum silicate ($Be_3Al_2(Si_2O_6)_3$). Beryl also contains other,

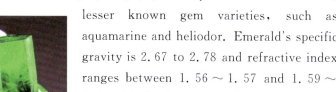

lesser known gem varieties, such as aquamarine and heliodor. Emerald's specific gravity is 2.67 to 2.78 and refractive index ranges between $1.56 \sim 1.57$ and $1.59 \sim 1.60$. It's hardness on Mohs scale is 7.5. Although this gemstone has a fairly good hardness, it is highly brittle. Pure beryl is white; emerald's green color is caused by chromium impurities (and occasionally by vanadium impurities).

It is the color of this gemstone that decides the emerald prices in the market. However, the cut, clarity and size of the gem are also deciding factors for the price of this gemstone. The deep green emerald color has always been the standard in grading all the other green colored gemstones. However there has been a disagreement among the experts in the field about the color of the emerald.

Traditionally any fine green beryl colored by chromium was called an emerald, although a lighter variety of this gem should be rightly called as a green beryl. But then again there is a difference as to what color should be considered light, even though there have been master stone sets that the gemological laboratories use to classify it from a green beryl.

So one would wonder what the most desirable color of the emerald would be. The answer to this is a bluish green to green and the trace elements may vary between chromium, vanadium and iron. The presence or absence of any of these will bring in a color difference in the emerald.

The most common treatment carried out on emeralds to enhance the stone is oiling. Many fillers are also used along with the oil. The natural emerald crystal is soaked in colored as well as colorless oil or resins for a particular amount of time. Many a times the oil may be heated, so that it seeps well in the fractures of the gems. This helps to fill in emerald fractures to make it look like a less included gem and enhance its color.

Although oiling is an accepted trade practice, treatment with colored oil, resin or any other filler should be disclosed before selling, as it can drastically bring down the price.

Emerald Gemstone had been synthesized in the laboratory several times by earlier scientists, but the first commercial synthetic emerald production was accomplished by Carroll Chatham around 1940. Even in 1961, a product developed by Johann Lechleitner of Austria was introduced by the Linde company as the "Linde Synthetic Emerald". Large crystals of over 1000 carats have been made by Chatham and later by Gilson. Cut gemstones of over 5 carats are commonly available in the market.

The cost of synthetic emerald or lab created emerald that is commonly found in the market is much lower than the natural ones. Synthetic emeralds may cost anywhere between $1 per carat to $150 per carat depending upon its quality and closeness to its natural counterpart. Synthetics are invariably very transparent and have the best green color. With the unaided eye it is very difficult to distinguish the synthetic emeralds from the natural ones. If in doubt, it is always better to get the stone tested at a reputed gem testing laboratory.

Some gemstones of lower value that could be confused with the natural green colored emerald are as follows: green beryl, demantoid garnet, green zircon, green sapphire, tourmaline, chalcedony, aventurine quartz, synthetic emerald and fluorite.

Part 8 Words and Expressions

craft[krɑːft]	n. 工艺,手艺
unique[juˈniːk]	adj. 唯一的,独特的
invoice[ˈɪnvɔɪs]	n. 发票,发货单,货物 v. 开发票,记清单
the classic nail inclusion	典型的钉状包裹体
account for	说明,占,解决,得分
lush[lʌʃ]	adj. 青葱的,味美的,豪华的,繁荣的
paradisiacal[ˌpærədɪˈsaɪəkəl]	adj. 天堂的,天堂似的,乐园的
landscape[ˈlændskeɪp]	n. 风景,山水画,地形,前景 v. 美化
Persian[ˈpɜːʃn]	n. 波斯人 adj. 波斯的,波斯人的
djebel[ˈdʒebəl]	n. 高山,山岭,山脉
tablet[ˈtæblɪt]	n. 写字板,书写板,碑,牌匾,拍纸簿,便笺簿,药片,小块
Zambia[ˈzæmbɪə]	n. 赞比亚(位于非洲)
Zimbabwe[zɪmˈbɑːbweɪ]	n. 津巴布韦(位于非洲)
Madagascar[mædəˈgæskə]	n. 马达加斯加岛(非洲岛国)
torture[ˈtɔːtʃə]	n. 折磨,痛苦,拷问,拷打
	vt. 拷问,曲解,折磨,使弯曲
Babylon[ˈbæbɪlən]	n. 巴比伦,奢华淫靡的城市
Goddess[ˈgɒdɪs]	n. 女神
Venus[ˈviːnəs]	n. 维纳斯,金星
represent[reprɪˈzent]	vt. 表现,描绘,声称,象征,回忆,再赠送,再上演
	vi. 提出异议
immortality[ˌɪmɔːˈtæləti]	n. 不朽,不朽的声名
gown[gaʊn]	n. 长袍,法衣,礼服,睡袍
beryllium aluminum silicate	铍铝硅酸盐
heliodor[ˈhiːlɪədɔː]	n. (产于非洲纳米比亚的)金绿柱石
impurity[ɪmˈpjʊərɪti]	n. 杂质,混杂物,不洁,不纯
vanadium[vəˈneɪdɪəm]	n. 钒(元素符号 V)
seep[siːp]	v. 渗出,渗漏
invariably[ɪnˈvɛərɪəblɪ]	adv. 不变地,总是
aventurine quartz	东陵石,砂金石英

Part 9 Check Your Understanding

Exercise 1 Answer the following questions

(1) Please tell gemological characteristics of emerald.
(2) Which countries was the emerald produced?
(3) Which factors may decide the price of emerald?
(4) Which mineral inclusions are commonly within ruby and sapphire?
(5) What is the main characteristic of synthetic emerald?

Exercise 2 Fill in the blanks

(1) The earliest known producing emeralds place is _____.
(2) The best emerald is produced in _____.
(3) Beryl contains known two gem varieties: _____ and _____.
(4) The most common treatment enhancing emeralds is _____.

Exercise 3 Translation

(1) The emerald is lucky for those who are born in the month of May. Hence it is also called as the May Birthstone.
(2) Emeralds are normally found to have visible inclusions. Thus it is an accepted practice in the market to sell emeralds with inclusions or flaws.
(3) Synthetic emeralds may cost anywhere between \$1 per carat to \$150 per carat depending upon its quality and closeness to its natural counterpart. Synthetics are invariably very transparent and have the best green color.
(4) 祖母绿的颜色常让人心醉,高质量的祖母绿比同等大小的钻石还要贵得多。
(5) 祖母绿和海蓝宝石因都是绿柱石的变种而具有非常相似的宝石学性质。
(6) 可以根据折射率值和相对密度值的差异来鉴别祖母绿和绿色碧玺。

Part 10 Self-Study Material

Price, Cost and Buying of Natural Emerald Gemstone

Unlike most of the other colored stones, the price of emerald gemstone mostly depends on its color, followed by the clarity, size and cut of the gem. Emerald color has always been the standard for grading all the other green-colored gems. Bluish green to green is the most desirable color in emeralds. Presence of yellow or too much of blue can bring down the price of the stone very much. The color should be evenly distributed throughout the stone, with no eye-visible zoning. Also, as the thumb rule, transparent stones are more expensive than included ones. A high transparency and a good color can command very high prices for an emerald.

To cite an example of how the color and clarity of a stone can make it so valuable, in 1998 a 5.16 carat untreated, slightly bluish green eye clean Colombian emerald was sold at twice the rate of a diamond the same size and of highest quality! But in the normal market emerald can cost anywhere from $10 per carat to $3000 per carat depending on whether it is small, medium or large or has an unmatched grade.

Emeralds are notorious for their flaws. Flawless stones are very uncommon, and are noted for their great value. Some people actually prefer an emerald with very minute flaws over a flawless emerald, as this proves authenticity of the stone. Many emerald flaws can be hidden by treating the emeralds with oil. Newer, more effective fracture-filling techniques are also practiced. Irradiation of some emerald gems is somewhat effective in removing certain flaws.

While buying a specific piece, all details of the emerald should be checked. If you are buying one to be set inside wedding jewelry, you may want to check out the heart shape designs. Make the choice based on various aspects mentioned above—color, cut, carat and clarity. You can also ensure that you have fancy shapes and include tear drop, oval and other shapes.

Unit 13　Quartz Gemstone

Part 1　Dialogue

Mary：Good afternoon, Madam, something for you?
Kate：Can you tell me what those gemstones are in the showcase?
Mary：Yes. They are fancy stones. Here are ruby, sapphire, tourmaline, aquamarine, amethyst, citrine, topaz, opal and etc.
Kate：I see. Let me look at these articles thoroughly before making decisions.
Mary：Go right ahead, please.
Kate：Wow, I like the red heart-shaped ring. What is the gem?
Mary：It's garnet with Italian style.
Kate：Gold? Silver? Platinum? Or K-gold jewelry?
Mary：It's silver. And it just costs ＄50.
Kate：Well. I'll get two like that. But what if there is any problem with them?
Mary：You can exchange for new ones within seven days of purchase.
Kate：Can I have my money back if I don't want to exchange?
Mary：Yes, no problem, Madam. I'll refund you in full.

本章音频

There is a warranty for them.
Kate: OK. Pack them for me. Thanks.

Part 2　Fill the Blanks Basing on Your Gemological Knowledge

Gemological characteristics of quartz:
chemical composition:_____　Mohs' hardness:_____
specific gravity:_____　refractive index:_____
familiar mineral inclusion:_____
main varieties of quartz:_____

Part 3　Link the Words and the Relevant Pictures

rose quartz
agate
bloodstone
jasper
sardonyx
orange chalcedony
star quartz
tiger's eye
rutilated quartz

Part 4　Useful Phrases

amethyst　紫晶
rock crystal　无色水晶
aventurine quartz　东陵石
smoky quartz　烟晶
citrine　黄晶
rose quartz　芙蓉石
rainbow quartz　虹彩水晶

bicolor quartz 双色水晶
colourless and transparent crystal 无色透明水晶
drusy quartz 石英晶簇
Australian chrysoprase 澳大利亚绿玉髓(澳玉)
rutilated quartz 金丝发晶
colored stone 彩色宝石
snake chain 蛇形链

Part 5　Some Sentences

(1) Bloodstone is an opaque, dark-green chalcedony with red spots caused by iron oxide.

(2) Some gemstones, notably garnet, topaz, peridot and tourmaline, belong to groups of minerals whose chemical compositions are related but which vary from specimen to specimen. In these gems, therefore, properties such as RI and SG differ from specimen to specimen.

(3) Jasper, a member of the chalcedony family (micro-crystalline quartz), opaque and fine grained. Jasper occurs in shades of red, yellow, green, greyish blue, brown and combinations of these. Jasper is partially organic.

(4) Tiger's eye quartz contains brown iron which produces its golden yellow color. Cabochon cut stones of this variety show the chatoyancy (small ray of light on the surface) that resembles the feline eye of a tiger. The most important deposit is in South Africa, though tiger's eye is also found in western Australia, Burma (Myanmar), India and California.

Part 6　Short Paragraph

The Purple Quartz Crystal—Amethyst

Amethyst, the most popular gem from the quartz family, once a precious and rare crystal gem of royalty and grace, has now became just another attractive and affordable gem today due to the abundance of the top quality amethyst crystals found all over the world. It is the birthstone for people born in the month of February and hence also called as the february birthstone. This fine gemstone gets its name from a Greek word "Amethustos" which means "not drunk" after the belief that the wearer of the gem will not suffer from excess consumption of alcohol. Amethyst geodes have beautiful purple pointed crystals which are loved by collectors all over the world.

History states that at one point this gemstone was so popular that it was as important and as expensive in the market as a ruby, emerald or a sapphire. It actually goes way back to 2500 B. C. in France where Neolithic man used it for decoration. Prior to that in 3100 B. C. , this gemstone was used in beads, jewelry and for seal making in Egypt and Greece.

All throughout Europe this gemstone had earned a royal appreciation till 1799. In 1799 a plentiful supply of high quality amethyst gemstones was discovered and mined in the Ural Mountains of Russia. These amethyst mines caused the price of the amethyst gemstones to decrease.

In Europe, in the middle ages, an amethyst gemstone was believed to protect the solders in the battlefield. The ancient Greeks believed that these gems had magical powers and these beautiful gemstones were used to make medicines.

Popular amethyst gemstone jewelry includes amethyst earrings, amethyst hoop earrings, and amethyst stud earrings. Because of their color and deep shape, amethyst gemstones of the purple variety are made into amethyst engagement rings and amethyst wedding rings.

Part 7 Text

Quartz Gemstone

Quartz is one of the most common minerals on earth and is well known in the gems world. Quartz is attractive and durable, as well as inexpensive. It can be cut and carved in many forms and sizes.

There are two main varieties of quartz, though they share the same chemical composition, silicon dioxide. Macrocrystalline quartz, includes stones like rock crystal (colorless), amethyst (violet), citrine (golden yellow), rose quartz (pink or peachy),

smoky quartz (brown), aventurine, quartz cat's eye and other varieties. The quartz is mostly transparent to translucent; Cryptocrystalline quartz, with microscopically small crystals, is known as agate, dendritic agate, fire agate, bloodstone, chrysoprase, prase, fossilized wood, jasper, carnelian, moss agate, onyx, sardonyx, and sard. Cryptocrystalline quartz is usually opaque or translucent.

The quartz gemstone has a hardness of 7 on the Mohs' scale. Its refractive index is 1.544~1.553 and it has birefringence of 0.009. Its crystal system is trigonal and it shows no cleavage. It is a transparent stone with a vitreous luster. Absorption spectrum and fluorescence differ within the varieties.

In gemstone culture, rock crystal and smoky quartz were once used for crystal balls that disclosed fortunetellers, witches or gypsy grandmas the future of their clients. Amethyst is the birthstone for those who are born in February, while Citrine is a birthstone for November.

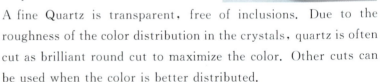

When buying quartz gem, the deep colors are the most valuable. A fine Quartz is transparent, free of inclusions. Due to the roughness of the color distribution in the crystals, quartz is often cut as brilliant round cut to maximize the color. Other cuts can be used when the color is better distributed.

While most varieties of transparent quartz are valued most when they show no inclusions, some are valued chiefly because of them! The most popular of these is known as rutilated quartz. Rutilated quartz is transparent rock crystal with golden needles of rutile arrayed in patterns inside it. Each pattern is different and some are breathtakingly beautiful. The inclusions are sometimes called Venus hair. Less well known is a variety called tourmaline quartz which, instead of golden rutile, has black or dark green tourmaline crystals.

Colorless quartz is always untreated. Colored stones can occasionally be enhanced in color by dying (as in the case of agate), irradiation (bombardment with low level radioactivity), or heating. Reliable gem dealers will always inform their customers about any kind of treatment.

quartz deposits:

Rock crystal: In the Alps, Brazil, Madagascar, U.S.A.
Smoky quartz: Brazil, Madagascar, Russia, Scotland, Switzerland, Ukraine.
Amethyst: Brazil, Bolivia, Canada, India, Madagascar, Mexico, Myanmar (Burma), Namibi-

a, Russia, Sri Lanka, United States (Arizona), Uruguay and Zambia.

Citrine: Argentina, Brazil, Madagascar, Namibia, Russia, Scotland, Spain, U.S.A.

Rose quartz: Brazil, India, Madagascar, Mozambique, Namibia, Sri Lanka, U.S.A.

Aventurine: Austria, Brazil, India, Russia, Tanzania.

Quartz cat's eye: Brazil, India, Sri Lanka.

Hawk's eye: Brazil, India, Sri Lanka.

Tiger's eye: Australia, India, Myanmar, Namibia, South Africa, Sri Lanka, U.S.A.

Part 8 Words and Expressions

showcase['ʃəukeɪs]	n.(商店或博物馆的玻璃)陈列橱,显示优点的东西
silver['sɪlvə]	n. 银,银子
exchange[ɪks'tʃeɪndʒ]	vt. 交换,调换,兑换,交易 n. 交换,调换,兑换,交易
purchase['pə:tʃəs]	vt. 买,购买 n. 买,购买
refund [ri:'fʌnd]	v. 退还,偿还 n. 归还,偿还额,退款
warranty['wɔrəntɪ]	n.(正当)理由,(合理)根据,授权,担保,保证,根据
variety[və'raɪətɪ]	n. 变化,多样性,种种,品种,种类
rose quartz	芙蓉石,玫瑰水晶
jasper[dʒæspə]	n. 碧玉
sardonyx['sɑ:dənɪks]	n. 红条纹玛瑙,缠丝玛瑙
star quartz	星光石英
rutilated quartz	金丝(含金红石的)发晶
opaque[əu'peɪk]	n. 不透明物 adj. 不透明的,不传热的,迟钝的
spot[spɔt]	n. 斑点,污点,地点,现场
	v. 玷污,弄脏,侦察 vt. 认出,发现
iron oxide	氧化铁
specimen['spesɪmən]	n. 范例,标本,样品,样本,待试验物
microcrystalline[,maɪkrəu'krɪstəlaɪn]	adj. 微晶的
microcrystalline quartz	微晶质石英
grained[greɪnd]	adj. 有木纹的
greyish['greɪʃ]	adj. 带灰的,略灰的
organic[ɔ:'gænɪk]	adj. 器官的,有机的,组织的,建制的
resemble[rɪ'zembl]	vt. 像,类似
purple['pə:pl]	adj. 紫色的
royalty['rɔɪəltɪ]	n. 皇室,王权
grace[greɪs]	n. 优美,雅致,优雅
affordable[ə'fɔ:dəbl]	adj. 提供得起的,供应得起的

consumption[kənˈsʌmpʃn]	n. 消费,消费量
alcohol[ˈælkəhɒl]	n. 酒精,酒
geode[ˈdʒiːəud]	n. 晶洞,异质晶簇
neolithic[niːəˈlɪθɪk]	adj. 新石器时代的
decoration[ˌdekəˈreɪʃn]	n. 装饰,装饰品
Egypt[ˈiːdʒɪpt]	n. 埃及
Greece[griːs]	n. 希腊
royal[ˈrɔɪəl]	adj. 王室的,皇家的,第一流的,高贵的
appreciation[əˌpriːʃiˈeɪʃn]	n. 感谢,感激,正确评价,欣赏,增值
Ural Mountains of Russia	俄罗斯的乌拉尔山脉
battlefield[ˈbætlfiːld]	n. 战场,沙场
magical[ˈmædʒɪkəl]	adj. 不可思议的
hoop[huːp]	n. 箍,铁环,戒指
hoop earrings	耳环
stud[stʌd]	n. 大头钉,纽扣
stud earrings	耳钉
carved[kɑːvɪd]	adj. 有雕刻的
silicon dioxide	二氧化硅
peachy[ˈpiːtʃɪ]	adj. 桃色的
quartz cat's eye	石英猫眼
cryptocrystalline[ˌkrɪptəuˈkrɪstəlɪn]	adj. 隐晶(质)的
cryptocrystalline quartz	隐晶(质)石英
microscopically[ˌmaɪkrəˈskɒpɪklɪ]	adv. 显微镜地,精微地
dendritic[denˈdrɪtɪk]	adj. 树枝状的
dendritic agate	树枝状玛瑙
prase[preɪz, preɪs]	n. 绿石英
fossilize[ˈfɒsɪlaɪz]	vt. 使成化石,使陈腐
	vi. 变成化石,搜集(或发掘)化石标本
fossilized wood	硅化木
carnelian[kɑːˈniːliən]	n. 红玉髓
moss agate	苔藓玛瑙
sard[sɑːd]	n. 肉红玉髓
vitreous luster	玻璃光泽
absorption spectrum	吸收光谱
fortuneteller[ˈfɔːtʃənˌtelə]	n. 算命者,占卜者
witch[wɪtʃ]	n. 巫婆,女巫,迷人的女子　vt. 施巫术,迷惑
gypsy[ˈdʒɪpsɪ]	n. 吉普赛人,吉普赛语
roughness[ˈrʌfnɪs]	n. 粗糙,粗暴,未加工
needles of rutile	针状金红石

Unit 13   Quartz Gemstone   • 129 •

breathtakingly['breθteɪkɪŋlɪ]   *adv.* 惊人地,惊险地
irradiation[ɪˌreɪdɪ'eɪʃn]   *n.* 放射,照射
bombardment[bɔm'bɑːdmənt]   *n.* 炮击,轰击
radioactivity[ˌreɪdɪəuæk'tɪvɪtɪ]   *n.* 放射能
inform[ɪn'fɔːm]   *v.* (~of/about) 通知,告诉,获悉,告知
Ukraine[juːˈkreɪn]   *n.* 乌克兰
Bolivia[bəˈlɪvɪə]   *n.* 玻利维亚(南美洲中部国家)
Uruguay['urugwaɪ]   *n.* 乌拉圭
Argentina[ˌɑːdʒən'tiːnə]   *n.* 阿根廷(南美洲东南部国家)
Spain[speɪn]   *n.* 西班牙(欧洲南部国家)
Mozambique[ˈməuzəm'biːk]   *n.* 莫桑比克(非洲南部国家)

Part 9   Check Your Understanding

Exercise 1   Answer the following questions

(1) Please tell gemological characteristics of quartz.
(2) Which factors may caused the price of the amethyst gemstones to decrease in 1779?
(3) Which factors may decide the price of quartz gem?
(4) Which countries were the amethyst produced?

Exercise 2   Fill in the blanks

(1) Bloodstone is an opaque, dark-green chalcedony with red spots caused by _____.
(2) Colored quartz gems can occasionally be enhanced in color by _____.
(3) The color of rutilated quartz is likely _____.
(4) February birthstone is _____.

Exercise 3   Translation

(1) Some gemstones, notably garnet, topaz, peridot and tourmaline, belong to groups of minerals whose chemical compositions are related but which vary from specimen to specimen.
(2) Jasper, a member of the chalcedony family (micro-crystalline quartz), opaque and fine grained. Jasper occurs in shades of red, yellow, green, greyish blue, brown and combinations of these.
(3) Macrocrystalline quartz, includes stones like rock crystal(colorless), amethyst(violet), citrine (golden yellow), rose quartz (pink or peachish), smoky quartz (brown), aventurine, quartz cat's eye and other varieties.
(4) 金发晶是因石英含有针状金红石,黑发晶是因石英含有针状电气石。
(5) 紫晶是石英族宝石中最名贵的品种。
(6) 在如今珠宝市场上,相对钻石和红宝石,一般石英族宝石的价格不算高。

Part 10 Self-Study Material

Chalcedony Variety of Quartz

Quartz that is formed not of one single crystal but a number of finely grained microcrystals is known as chalcedony. The variety of chalcedonies is even greater than that of transparent quartz, including cryptocrystalline quartz with patterns as well as a wide range of solid colours. Agates are banded. Bloodstone has red spots on a green background. Moss agate has a plant-like pattern. Jasper sometimes looks like a landscape painting. Another staple of the jewelry industry is black onyx, chalcedony quartz which owes its even black colour to an ancient dyeing process that is still used today. Carnelian, another chalcedony valued in the ancient world, has a vivid brownish orange colour and clear translucency that makes it popular for signet rings and seals. Chrysoprase, a bright, apple-green, translucent chalcedony, is the most valued. Today, chrysoprase is found mostly in Australia. Unlike most other green stones, which owe their color to chromium or vanadium, chrysoprase derives its color from nickel. Its bright even color and texture lend themselves well to beads, cabochons, and carvings.

Chalcedony has a hardness of 6.5 on the Mohs' scale and a specific gravity of 2.58 to 2.64 with a refractive index of 1.544 to 1.533. It has a hexagonal structure with conchoidal fracture and no cleavage. It can be transparent and opaque with a dull, waxy, vitreous luster. It is also piezoelectric.

Due to the wide variety of the stone available in the market, chalcedony is priced rather inexpensively, but stones with striking features, good color and clarity command high prices. Chalcedony occurs in many countries round the world, but the finest specimens come from India, Burma, Brazil, Mexico, USA, Madagascar, Austria, Iceland, New Zealand, Russia and Britain.

Unit 14　Pearl

Part 1　Dialogue

本章音频

Mary: Good morning, welcome to our shop, Madam. Do you have any particular types in mind?

Kate: Well, yes. I'd like to give a set of jewelry for my mother.

Mary: Please come over here. Something smart? Something for special occasion?

Kate: I've no idea, could you show me something elegant but inexpensive?

Mary: Pearl is better for your mother, I suppose. Do you have a budget?

Kate: No, I know little about the jewelry price.

Mary: I recommend this set including ear pendants, necklace and hand chain.

Kate: It looks beautiful, natural or cultured?

Mary: Cultured seawater pearls, from Japan. Many customers have high comments on them.

Kate: What is the price?

Mary: $4800.

Kate: Give me the real price, boss.

Mary: I'm sorry. We hold a one-price policy.

Kate: Oh, That's too expensive, I'm afraid. What a pity!

Mary: In fact, the price is reasonable for the quality of this article.

Kate: Well, I'll take around.
Mary: Sorry, make a decision before you come back.
Kate: Thanks.

Part 2　Fill the Blanks Basing on Your Gemological Knowledge

Gemological characteristics of pearl:
chemical composition: _____　　Mohs' hardness: _____
specific gravity: _____　　refractive index: _____
structure: _____
types of pearls: _____

Part 3　Link the Words and the Relevant Pictures

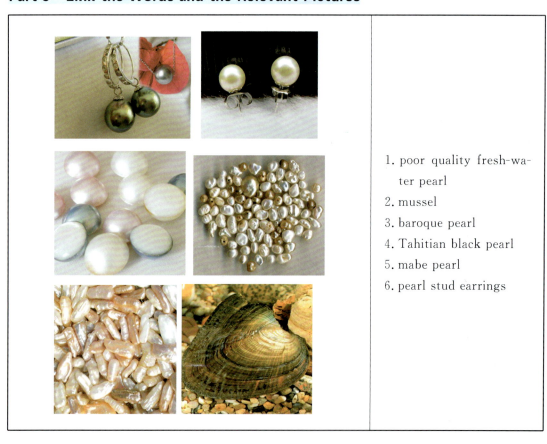

1. poor quality fresh-water pearl
2. mussel
3. baroque pearl
4. Tahitian black pearl
5. mabe pearl
6. pearl stud earrings

Part 4　Useful Phrases

saltwater cultured pearls　海水养殖珍珠
mother of pearl　珠母,珠母质阿古屋贝珍珠
akoya pearls　日本珠(akoya 是一种蚝的名称)
night-luminous pearl　夜明珠
Tahitian pearl　大溪地珍珠（一般为黑色）

tortoise shell　玳瑁（龟甲）
drilled pearl　有孔珍珠
dust/seed pearl　小珍珠
baroque stone　原形宝石
button pearl　纽扣珍珠
cyst pearl　囊珍珠
dead pearl　无光珍珠
hyriopsis cumingii　三角帆蚌
baroque pearl(crop pearl)　（巴洛克）异形珍珠
blister pearl（Mabe pearl）　附贝（或贝附）（马白）珍珠

Part 5　Some Sentences

（1）Natural pearls are produced entirely by natural processes whereas cultured pearls are deliberately initiated and farmed by man.

（2）The most important pearl-producing molluscs are saltwater oysters and freshwater mussels.

（3）Pearls are categorized as organic gem material and are amongst the oldest of precious gems. In history pearls have been very valuable, second only to the diamond.

（4）Pearls are relatively soft and can be damaged through scratching and exposure to heat and chemicals. When taken care of properly, however pearls will last a lifetime and beyond.

（5）Cultured pearls are relatively inexpensive as compared to its natural counterpart. Good quality 3mm sized strand of cultured pearls should cost anywhere between $10 per gram to $25 per gram. The cost mainly depends upon the shape, luster and its surface finish.

Part 6　Short Paragraph

Natural Pearls

Composition: natural pearls consist of over 80% calcium carbonate, 10% to 14% conchin (or conchiolin) and 2% to 4% water. Calcium carbonate is present as the orthorhombic form, aragonite. Calcite may be present in some freshwater pearls.

Specific Gravity: 2.60～2.78.

Refractive index: 1.530～1.685.

Hardness on Mohs': 3.5~4.

Structure: Natural pearls possess a concentric structure consisting of a microscopically small nucleus, surrounded by layers of calcium carbonate, variety aragonite, cemented together by conchin. Each layer represents one season's growth.

Within the carbonate layers, the c-axes of the crystals are arranged radially. This feature has important implications for gem testing. The aragonite crystals are platy and overlap, rather like tiles on a roof.

Luster: The unique luster of pearl is a complex effect, quite different from the luster of surfaces of polished gemstones. It is produced partly by the surface reflection of light and partly by a subtle iridescence. This iridescence is caused by:

a. diffraction at the edges of overlapping platy crystal of aragonite;

b. interference when light penetrates some of the thin plates, and is then reflected back to the surface.

The resulting combined reflection, interference and diffraction effect is known as the orient of pearl. This is not a scientifically definable term; it is discernible as a subtle effect, distinct from bodycolor. Appreciation of its quality is dependent upon the experience of the viewer.

Color: Body colors can be broadly classified as white, pink, yellow, grey, bronze, green, mauve and black. In addition, the orient itself may show delicate but predominating tints of pink, yellow, green or blue. Many theories have been advanced regarding the causes of color in grey and black pearls. These include the presence of certain minerals and proteins in the water in which the oysters have developed, and the existence of a conchin-rich layer near to the surface of the pearls.

Shape: Ideally spherical, but many pearls depart from this shape. Pear (or drop) shaped pearls are also highly prized. Flattened spheres are called bouton or "button" pearls. Highly irregularly shaped pearls are called baroque pearls.

The limited availability of natural pearls has provoked man to experiment in stimulating the pearl from oyster to produce pearl with human help. The Chinese have been trying to culture pearls from the 13th century.

Part 7   Text

Pearl

The pearl, the queen of the gem, is favored by people all through the human history. It can be classified as either natural or cultured pearl. The creation of a natural pearl begins when a foreign object enters the shell of an oyster without any assistance from a human being. This object irritates the oyster and to rid itself of this irritant, the oyster coats this object with layers of a smooth substance called nacre. Over the years, the layers of nacre get thicker and thicker, eventually creating a beautiful and lustrous pearl.

In order to create a cultured pearl, it is necessary to imitate nature. The oyster's shell is carefully opened and an irritant is placed inside, thus stimulating the production of nacre. At this point, nature takes over and in due time a lustrous pearl is born. Another distinction among pearls is between seawater pearls and those grown in freshwater. The irritant for a

freshwater mussel is a slice of mantle while a seawater oyster welcomes a kind of small ball ground from shell, called nucleus. The production of cultured pearls has been accessible to man's will. The jewelers turn to cultured pearls since they are highly competitive to their natural counterparts in either quality or price. Along with over harvest and water pollution, the natural pearls are rarely found except in auctions or antique stores.

When selecting the pearls of your choice, there are five characteristics to consider: luster, texture, shape, size and color.

Luster: Luster is the beautiful shine of the pearl created by the many coatings of nacre. The more coatings of nacre the pearl has, the more lustrous it is. It is luster that is the most important factor in the grading of pearls.

Texture: Since pearls are a natural creation, it is almost impossible to find a pearl without any imperfections. The smaller the imperfections are, the higher the quality of the pearl.

Shape: Pearls are seldom perfectly round or even nearly round; more often than not they are baroque. The rounder the pearl, the more expensive it will be.

Size: Pearls are measured at their diameter with the sizes ranging from 1mm to 20mm. As the pearls increase in size, so does their value.

Color: Although it is possible to find pearls of blue, black, gold, pink and orange colors, the most common and desirable pearls are white and cream.

Occurrence: The most important pearl fisheries in the past were in the Persian Gulf and the Gulf of Manaar (between Sri Lanka and India). Fishing still continues in these waters, but to a much reduced extent owing to pollution and changed economic circumstances. Off the northern coasts of Australia, pearls are recovered in the course of fishing for mother of pearl shell. Other areas include the Pacific Islands, the Red Sea, Venezuela and the Gulf of Mexico.

The bulk of cultured pearl farming is centred in China and Japan. Production is also carried out in Australia, Myanmar, Philippines, Tahiti, Thailand and other countries.

Part 8 Words and Expressions

elegant['elɪgənt]　　　　　　　　　　adj. 文雅的,端庄的,雅致的,上品的,第一流的
inexpensive[ˌɪnɪks'pensɪv]　　　　　adj. 便宜的,不贵重的
budget['bʌdʒɪt]　　　　　　　　　　n. 预算　vi. 做预算,编入预算
seawater pearl　　　　　　　　　　海水珍珠
deliberately[dɪ'lɪbərətlɪ]　　　　　　adv. 故意地
initiate[ɪ'nɪʃɪeɪt]　　　　　　　　　　vt. 开始,发动,传授　v. 开始,发起
mollusc['mɒləsk]　　　　　　　　　　n. 软体动物
categorize['kætəgəraɪz]　　　　　　v. 加以类别,分类
strand[strænd]　　　　　　　　　　n. 绳、线之一股,线,绳,串,海滨,河岸
finish['fɪnɪʃ]　　　　　　　　　　　　n. 完成,结束,磨光,完美
calcium['kælsɪəm]　　　　　　　　　n. 钙(元素符号 Ca)
carbonate['kɑːbəneɪt]　　　　　　　n. 碳酸盐
conchiolin[kɒŋ'kaɪəlɪn]　　　　　　n. 壳质,介壳质
orthorhombic[ˌɔːθə'rɒmbɪk]　　　　adj. 斜方晶系的
aragonite[ə'rægənaɪt]　　　　　　　n. 文石
concentric[kən'sentrɪk]　　　　　　adj. 同心的,同轴的
microscopically[ˌmaɪkrə'skɒpɪklɪ]　adv. 用显微镜,通过显微镜
nucleus['njuːklɪəs]　　　　　　　　　n. 核子,中心(nuclear 的复数)
radially['reɪdjəlɪ]　　　　　　　　　adv. 放射状地
implication[ˌɪmplɪ'keɪʃn]　　　　　n. 暗示
platy['pleɪtɪ]　　　　　　　　　　　　n. 裂成平坦薄片的,板状的,扁平状的
tile[taɪl]　　　　　　　　　　　　　　n. 瓦片,瓷砖
roof[ruːf]　　　　　　　　　　　　　n. 屋顶,房顶,顶
subtle['sʌtl]　　　　　　　　　　　　adj. 微妙的

iridescence[ˌɪrɪˈdesəns]	n. 彩虹色
diffraction[dɪˈfrækʃn]	n. 衍射,宽龟裂状的
overlap[ˌəʊvəˈlæp]	v.(与……)交迭
interference[ˌɪntəˈfɪərəns]	n. 冲突,干涉
penetrate[ˈpenɪtreɪt]	vt. 穿透,渗透,看穿,洞察
	vi. 刺入,看穿,渗透,弥漫
definable[dɪˈfaɪnəbl]	adj. 可定义的
body color	n. 体色
mauve[məʊv]	n. 紫红色
predominant[prɪˈdɒmɪnənt]	vt. 掌握,控制,支配
	vi. 统治,成为主流,支配,占优势
tint[tɪnt]	n. 色彩
protein[ˈprəʊtiːn]	n. 蛋白质 adj. 蛋白质的
spherical[ˈsferɪkl]	adj. 球的,球形的
flatten[ˈflætn]	vi. 变平,变单调
	vt. 使平,变平,打倒,使失去光泽
baroque[bəˈrɒk]	adj. 巴洛克风格的
availability[əˌveɪləˈbɪlətɪ]	n. 可用性,有效性,实用性
provoke[prəˈvəʊk]	vt. 激怒,挑拨,煽动,惹起,驱使
stimulate[ˈstɪmjuleɪt]	vt. 刺激,激励 vi. 起促进作用
irritant[ˈɪrɪtənt]	n. 刺激物 adj. 刺激的
nacre[ˈneɪkə]	n. 珠母贝,真珠质
lustrous[ˈlʌstrəs]	adj. 有光泽的
competitive[kəmˈpetətɪv]	adj. 竞争的
texture[ˈtekstʃə]	n.(织品的)质地,(木材,岩石等的)纹理,(皮肤)肌理,(文艺作品)结构
imperfection[ˌɪmpəˈfekʃn]	n. 不完整性,非理想性
diameter[daɪˈæmɪtə]	n. 直径
fishery[ˈfɪʃərɪ]	n. 渔业,水产业,渔场,养鱼术
Persian[ˈpɜːʃn]	n. 波斯人[语] adj. 波斯的,波斯人[语]的
Venezuela[ˌveneˈzweɪlə]	n. 委内瑞拉(南美洲北部国家)
Gulf of Mexico	墨西哥海湾
bulk[bʌlk]	n. 大小,体积,大批,大多数,散装
	vt. 显得大,显得重要
Philippines[ˈfɪlɪpiːnz]	n. 菲律宾共和国,菲律宾群岛
Tahiti[təˈhiːtɪ]	n. 塔希提岛(位于南太平洋,法属波利西亚的经济活动中心)

Part 9 Check Your Understanding

Exercise 1   Answer the following questions

(1) How was the seawater pearls born?
(2) How was the pearl body color broadly classified?
(3) What should be thought over when we select the pearls?
(4) Which countries were the cultured pearl produced?

Exercise 2   Fill in the blanks

(1) Pearls consist of over 80% _____.
(2) The unique luster of pearl is produced by _____.
(3) The creation of a natural pearl begins when _____ enters the shell of an oyster.
(4) Pearls are measured at their _____ in the jewelry market.

Exercise 3   Translation

(1) The unique lustre of pearl is a complex effect, quite different from the luster of surfaces of polished gemstones. It is produced partly by the surface reflection of light and partly by a subtle iridescence.
(2) The pearl, the queen of the gem, is favored by people all through the human history. It can be classified as either natural or cultured pearl.
(3) When selecting the pearls of your choice, there are five characteristics to consider: luster, texture, shape, size and color.
(4) 珍珠属有机宝石,自古以来就是一种名贵宝石。
(5) 珍珠按照成因可分为天然珍珠和养殖珍珠,按水域又可分为淡水珍珠和海水珍珠。
(6) 中国是世界上淡水珍珠的主要出产国。

Part 10 Self-Study Material

The Care and Enhancement for Pearl

Care for Pearls

Natural Pearls are very delicate gems and can be easily damaged by the surrounding conditions. It is extremely important to care for pearl jewelry. With proper care the gems will last for centuries.

(1) When pearls are worn very often and close to the skin they get eroded or damaged with contact with even the mildest of acids given out by the skin. Hence the use of pearl jewelry very often in humid places is not advisable.

(2) Pearl jewelry, like pearl earrings, pearl rings etc. should not be kept in cotton wool as it contains small amounts of acids that may damage the pearl in the long run.

(3) String of pearls or pearl jewelry should not be kept in polythene bags as there isn't enough moisture for the pearls in these bags, which will create a water loss and damage the

outer surface of the pearls.

(4) The best way of storing pearls would be to keep them well wrapped in white linen cloth or pure silk.

(5) Cosmetics and perfumes must never come in contact with pearls as the acids and chemicals most certainly will damage them.

(6) Restringing of pearls that are very often used is a very good idea as many a times the string may absorb the perfume of cosmetics used and in turn damage the pearl. The string used (if silk) by itself may wear out or may break. It is advisable to restring regularly used pearls once every six months.

(7) The best way of stringing pears is to have a knot at the end of each pearl so that in case of breakage of the string only one pearl is lost.

Enhancement for Pearls

Staining: Poorly colored pearls may be stained black by soaking in a solution of silver nitrate, followed by exposure to light. The chemical decomposes depositing a subsurface layer of metallic silver, giving a grey on black appearance. Pale tints may be induced by soaking in suitable dyes.

Irradiation: This may produce a grey color.

Bleaching: This may be accomplished by soaking in a mild bleaching agent, such as dilute hydrogen peroxide. This treatment may damage pearls.

Skinning: Skinning pearls involves the removal of successive layers of nacre, in the hope of improving their shape, color or luster. It is a process which requires the highest degree of specialist skill, and is extremely risky. Historically, this was done by "pearl doctors".

Unit 15　Jadeite Jade

Part 1　Dialogue

Mary：Can I help you in any way? What do you wish to buy?

Kate：Yes. I want to buy a jadeite jade ornament.

Mary：What type of jadeite jade do you like best? Bangle, jade pendant, ring or eardrop?

Kate：I'm not quite certain. Show me some high-quality goods.

本章音频

Mary：All right，Madam. We have some really old jade from Burma.

Kate：The bangle is very beautiful.

Mary：It has been reserved，Madam.

Kate：Can I try this butterfly-shape jadeite brooch?

Mary：Oh，It fits you very well. The emerald green shines your appearance.

Mary：OK? It is expensive but it's worth every penny of it.

Kate：Let me think about.

Mary：Don't hesitate，Madam. I don't think we have much left. It's the latest fashion，done by master craftsman.

Kate：Can you give me a discount?

Mary：Sorry. The price is fixed. I guarantee the quality. I don't think we have many left.

Kate: All right, I just take this one.

Part 2　Fill the Blanks Basing on Your Gemological Knowledge

Gemological characteristics of jadeite:

chemical composition: _____　Mohs' hardness: _____

specific gravity: _____　refractive index: _____　structure: _____

luster: _____　treatment: _____

Part 3　Link the Words and the Relevant Pictures

glassy species
golden-thread species
oil-green species
violaceous species
icy species
sooty species

Part 4　Useful Phrases

actual or natural jadeite jade/A-grade jadeite　A 货翡翠
nleached jadeite jade/B-grade jadeite　B 货翡翠
coloration jadeite jade/C-grade jadeite　C 货翡翠
jade carving (sculpture)　玉雕
master carver　玉雕大师
jade carver　玉雕工人
exquisite jade　翡翠精品
jade luck　翡翠缘
old-pit glassy species　老坑玻璃种
icy species　冰种
lotus flower species　芙蓉种
golden-thread species　金丝种

dry-green species　干青种
green in white base　白地青
spotted or banded green　花青种
spotted blue　飘兰花种
oil-green species　油青种
bean species　豆种
horse-tooth species　马牙种
htelongsein jadeite jade　铁龙生翡翠
glutinous rice species　糯米种
jadeite jade stone gambling　翡翠赌石
jadeite jade bracelet, Guanyin pendant and Buddha　翡翠手镯、观音、佛
Feicui embedded in this necklace　镶嵌在项链上的翡翠
studies of genesis for primary deposit of Burma jadeite jade　缅甸翡翠原生矿床成因研究
genesis of jadeite jade ore from Burma　缅甸翡翠矿石的成因探讨
mutton fat jade　羊脂玉
jade article appraisal services　玉器评估服务
moss agate　苔藓玛瑙

Part 5　Some Sentences

(1) The term "jade" is applied to two different rocks that are made up of different silicate minerals—jadeite and nephrite.

(2) While most gemstones today are sold and evaluated in terms of their carat weight, jadeite jade is usually sold by the piece.

(3) As with other precious stones, there have

been many attempts to enhance and even create synthetic jadeite jade. Currently there is no commercially available synthetic jadeite jade in the gemstone market.

(4) The traditional jadeite jade identification methods mainly relied on refractometer, gemological microscope, Chelsea color filter and spectroscope, all these methods identified bleached jadeite jade(so called B jade) is very difficult.

(5) Jadeite jade(Fei Cui) in China was originally a bird name. "Fei" is a red feather birds, "Cui" is a kind of green feathers of the bird. Then gradually the word merger specifically refers to a blue-green with a brown bird on the surface.

Part 6　Short Paragraph

Chinese Jade Culture

Jade culture is deep and rich in China as the stone symbolizes beauty, nobility, perfection, constancy, power and immortality. The Chinese love jade not only for its beauty, but also for its culture, symbolic meaning and humanity. As Confucius said, there are eleven virtues in jade. Thus jade is really special in Chinese culture, as the Chinese saying goes "Gold has a value; jade is invaluable".

The history of jade is as long as the Chinese civilization. Archaeologists have found jade objects from the early Neolithic period, represented by the Hemudu culture in Zhejiang Province, and from the middle and late Neolithic period, represented by the Hongshan culture along the Lao River, the Longshan culture along the Yellow River, and the Liangzhu culture in the Taihu Lake region.

Because jade stands for beauty, grace and purity, it has been used in many Chinese idioms or phrases to denote beautiful things or people, such as Yu Jie Bing Qing (pure and noble), Ting Ting Yu Li (fair, slim and graceful) and Yu Nü (beautiful girl). The Chinese character Yu is often used in Chinese names.

Jade has been made into sacrificial vessel, tools, ornaments, utensils and many other items. Ancient music instruments were made out of jade, such as jade flute, yuxiao (a vertical jade flute) and jade chime. Jade was also mysterious to the Chinese in the ancient time, so wares were popular as sacrificial vessels and were often buried with the dead. To preserve the body of the dead, Liu Sheng, the ruler of the Zhongshan State (113 BC) was buried in a jade burial suit composed of 2498 pieces, sewn together with gold thread.

Part 7　Text

Jadeite jade

Gem quality jadeite jade occurs as a polycrystalline rock. It is one of the two materials known as jade, the other being nephrite. A combination of great toughness with a subtle

beauty, makes jadeite jade ideally suited to carving, or to fashioning as beads and cabochons. Gemological characteristics of jadeite are listed below.

(1) Chemical composition: NaAl$(SiO_3)_2$ Sodium aluminium silicate.

(2) Crystal system: Monoclinic.

(3) Habit & occurrence: the jadeite jade used for carvings and gemstones is polycrystalline. It occurs as a metamorphic rock and is also found in alluvial deposits as boulders and pebbles. It has a tough, granular, interlocking structure. Within the grains a fibrous texture may be visible with a lens.

(4) Hardness: 7 approximately, with slight directional differences.

(5) Specific gravity: 3.30 to 3.36.

(6) Colours: White, shades of mauve, violet, red, orange, yellow, brown, pale to deep emerald green, typically mottled green and white, deep green to black. Jadeite jade boulders may show a mottling of several colours; weathering of the outer surface produces a brown "skin". The most highly prized variety is the translucent to almost transparent emerald green variety known as "Imperial Jade".

(7) Absorption Spectrum: The absorption of jadeite jade is characterized by a strong line in the blue/violet which may be accompanied by weaker bands in the blue. Emerald green jadeite jade is colored by chromium and has a typical chromium absorption spectrum showing a doublet in the red, sometimes with a line in the blue. All other green and brown colors are caused by iron. Mauve jadeite jade is colored by manganese.

(8) Luster: Greasy to vitreous.

(9) Transparency: Transparent (very rare), translucent to opaque.

(10) Refractive index: 1.65 to 1.67. Usually only a vague shadow edge is seen on the refractometer at approximately 1.66.

(11) Birefringence: The random orientation of the doubly refracting crystals results in translucent pieces appearing light in all positions between crossed polarizers.

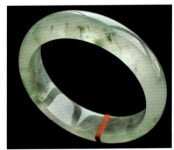

(12) Localities: The main commercial source is Myanmar, but it also occurs in Japan,

California USA, Guatemala and Russia.

(13) Fashioning: Cabochons, beads, carvings and small carved pieces for jewelry. A characteristic uneven polish or "orange-peel" effect is visible with a lens on older pieces, and on those items not polished with a diamond abrasive. This uneven polish is due to the differential hardness of the randomly orientated crystal grains in the rock. Polishing with a diamond abrasive produces a smooth and more highly lustrous (vitreous) surface.

(14) Treatment: Blanch, staining, filling and other treatments are common practice. Jadeite jade imitation s include translucent emerald, chalcedony, bowenite & other serpentines, glass & plastics.

Part 8 Words and Expressions

bangle['bæŋgl]	n. 手镯
eardrop['ɪədrɔp]	n. 耳坠, 耳饰
reserve[rɪ'zəːv]	n. 储备(物), 储藏量, 预备队
	vt. 储备, 保存, 保留, 预定, 预约
butterfly['bʌtəflaɪ]	n. 蝴蝶, 蝶泳
butterfly-shape	蝴蝶型
penny['penɪ]	n. 便士, 〈美〉分
vitreous['vɪtrɪəs]	adj. 玻璃质的
glassy species	玻璃种(翡翠)
violaceous species	紫罗兰种(翡翠)
sooty species	乌鸡种(翡翠)
silicate['sɪlɪkɪt]	n. 硅酸盐
commercially[kə'məːʃəli]	adv. 商业上, 通商上
available[ə'veɪləbl]	adj. 可用到的, 可利用的, 有用的, 有空的, 接受探访的
chelsea color filter	查尔斯滤色镜
spectroscope['spektrəskəup]	n. 分光镜
merger['məːdʒə]	n. 合并, 归并
constancy['kɔnstənsɪ]	n. 不屈不挠, 坚定不移, 恒久不变的状态或性质
immortality[ˌɪmɔː'tælɪtɪ]	n. 不朽, 不朽的声名
Confucius[kən'fjuːʃəs]	n. 孔子(公元前551—479年, 中国春秋末期思想家、政治家、教育家, 儒家的创始者)
archaeologist[ˌɑːkɪ'ɔlədʒɪst]	n. 考古学家
Neolithic[ˌniːə'lɪθɪk]	adj. 新石器时代的
idiom['ɪdɪəm]	n. 成语, 方言, 土语, 习惯用语

denote[dɪ'nəʊt]	vt. 指示,表示
sacrificial[ˌsækrɪ'fɪʃl]	adj. 牺牲的
vessel['vesl]	n. 船,容器,器皿,脉管,导管
utensil[juː(:)'tensl]	n. 器具
flute[fluːt]	n. 长笛,笛状物,凹槽
chime[tʃaɪm]	n. 一套发谐音的钟(尤指教堂内的),和谐
	vi. 鸣,打,和谐　vt. 敲出和谐的声音,打钟报时
ancient['eɪnʃənt]	adj. 远古的,旧的
polycrystalline[ˌpɒlɪ'krɪstəlaɪn]	adj. 多晶的
sodium aluminium silicate	钠铝硅酸盐
monoclinic[ˌmɒnə'klɪnɪk]	adj. 单斜(晶系)的
metamorphic[ˌmetə'mɔːfɪk]	adj. 变形的,变质的,改变结构的
alluvial[ə'luːvɪəl]	adj. 冲积的,淤积的
boulder['bəʊldə]	n. 大石头,漂石
pebble['pebl]	n. 小圆石,小鹅卵石
granular['grænjʊlə]	adj. 由小粒而成的,粒状的
interlocking structure	交织结构
grain[greɪn]	n. 谷物,谷类,谷粒,细粒,颗粒,粮食
fibrous['faɪbrəs]	adj. 含纤维的,纤维性的
mauve[məʊv]	n. 紫红色　adj. 紫红色的
mottled['mɒtld]	adj. 杂色的,斑驳的
mottling['mɒtlɪŋ]	斑纹,色斑
translucent[trænsˈluːsənt]	adj. 半透明的
iron['aɪən]	n. 铁,熨斗,坚强,烙铁,镣铐
	vt. 烫平,熨,用铁装备　vi. 烫衣服
greasy['griːzɪ]	adj. 多脂的,油脂的,油污的
vague[veɪg]	adj. 含糊的,不清楚的,茫然的,暧昧的
approximately[ə'prɒksɪmətlɪ]	adv. 近似地,大约
random['rændəm]	n. 随意,任意　adj. 任意的,随便的,胡乱的　adv. 胡乱地
orientation[ˌɔː(:)rɪen'teɪʃn]	n. 方向,方位,定位,倾向性,向东方
polarizer['pəʊləraɪzə]	n. 偏光器,起偏镜
crossed polarizers	正交偏光
Guatemala[ˌgwætɪ'mɑːlə]	n. 危地马拉(拉丁美洲),危地马拉人
abrasive[ə'breɪsɪv]	n. 研磨剂　adj. 研磨的
blanch[blɑːntʃ]	v. 漂白,使变白,遮断阳光,发白
staining['steɪnɪŋ]	n. 着色
filling['fɪlɪŋ]	n. 填补物,饼馅,纬纱,填充,供应
simulant['sɪmjʊlənt]	adj. 拟态的,模拟的
chalcedony[kæl'sedənɪ]	n. 玉髓

bowenite[ˈbəʊnaɪt]	鲍文玉（蛇纹石玉），国外通常将质地较纯、半透明的蛇纹石块状集合体或蛇纹岩称为鲍文玉
serpentine[ˈsɜːpəntaɪn]	n. 蛇纹石

Part 9 Check Your Understanding

Exercise 1 Answer the following questions

(1) How was the Feicui for jadeite jade derived from?

(2) Please illustrate cradlelands of China jade culture.

(3) How was the jadeite jade cut?

(4) Which countries the jadeite jade produced in?

Exercise 2 Fill in the blanks

(1) In jade culture, the stone symbolizes _____.

(2) Emerald green jadeite jade is colored by _____.

(3) Brown jadeite jade is colored by _____. Mauve jadeite jade is colored by _____.

(4) C-grade jadeite jade was treated by _____.

Exercise 3 Translation

(1) While most gemstones today are sold and evaluated in terms of their carat weight, jade is usually sold by the piece.

(2) Jade culture is deep and rich in China as the stone symbolizes beauty, nobility, perfection, constancy, power, and immortality.

(3) 翡翠和软玉都是档次比较高的玉石品种。

(4) 选购翡翠时，我们要考虑其颜色、质地、透明度和工艺等因素。

(5) 翡翠一般呈玻璃光泽，折射率约为 1.66，相对密度约为 3.33.

Part 10 Self-Study Material 1

Valuation of Jadeite Jade

Let's imagine, if you want the best diamond, absolutely colorless, ideal cut, flawless and in big size, that will end up with a sky-high price. The same thing happens in jadeite jade. If you want jadeite jade with imperial green color, glassy variety, no flaw or crack and perfect carved, you will meet an extremely high price, too. So, you'd better figure out your budget, and what character you want most, a vivid color, or good transparency, or great craftsmanship, before shopping.

Chinese have an old saying that "There is a price for gold, but not one for jade". It tells the truth. There are quoted prices lists for diamonds and most of other precious gemstones, which can help businessmen to evaluate their prices. Jadeite jade is the only gemstone for which regular prices are not quoted by the major international gemstone trading firms. It is because it's difficult to quote a price for jadeite jade. Jadeite jade differs from most other gems as it is not only polycrystalline aggregates but also not monocrystalline mineral. It is composed of many types of minerals. What are the principles of jadeite jade appraisal?

Color is the most important for determining value. Naked eyes are good tools to judge it. The best light source to view the stone for appraisal is daylight.

Transparency is very important for the appreciation of jadeite jade. Sometimes this quality is even more important than color. We always use variety to mean the transparency. Glassy variety is very transparent; Icy to Glassy variety is transparent; Icy variety is sub-transparent; Rice to icy variety is translucent; Pea variety is opaque. Of course, there are not only these several varieties in the world of jadeite jade. The Chinese have an old saying that "There are ten thousand varieties of jade". We will have a separate article to talk about varieties.

There are mainly two kinds of jadeite jade cutting: polished cabochon ones and carved ones. When examining polished cabochon ones, the following factors should be considered: the shape, symmetry, width and height proportions. To carved ones, the modeling, their form, contours and lines should be considered.

Texture, clarity, cracks and volume are also the factors of impacting the price of the jadeite jade.

Part 11 Self-Study Material 2

Nephrite Jade

Although "jade" has been in use for a variety of utilitarian and artistic purposes for over 7000 years, it was only in 1863 that a gemological distinction was made between the two different species commonly given this name. Jadeite, an aggregate of granular pyroxenes, actually is not related to nephrite, an aggregate of fibrous amphiboles. The fact that they occur in the same color and translucency range, are both incredibly tough, and were traditionally used for the same purposes, along with their superficially similar appearance has led to the odd consequence of having two quite different gems with the same name.

Even though marketers, jewelers and the public continue to refer to both gems as jade, more properly the species should be used as, or at least included in, the name: so either nephrite (or nephrite jade) or jadeite (or jadeite jade) is the preferred terminology.

Nephrite is, then, a calcium-magnesium silicate that varies from translucent to opaque, and from shades of green, through browns and yellows to greys and near whites as it varies in the proportion of the amphibole minerals in its makeup. The darker pieces are mostly made up of iron-rich (up to 5% iron content) actinolites, the lighter pieces contain more of the magnesium-rich tremolites. Pieces may be mottled or banded in color, and black inclusions are common. Typically the iron induced green colors of nephrite are dulled somewhat by brown tones in comparison to the more highly saturated chromium derived hues of green jadeite.

As nephrite jade is, by definition, a mixture of amphiboles with an interlocking microcrystalline structure, pure actinolite or pure tremolite minerals, therefore are not jade. Actinolite sometimes occurs in a chatoyant form, though, which is often sold under the misnomer "cat's eye jade".

Nephrite is mined in many locales, ranging from New Zealand, Siberia and South Korea, to the USA (Wyoming and California primarily), but the largest deposits, by far, come from British Columbia. These Canadian sites often yield huge boulders, frequently covered with a brown rind of oxidized iron. The finest of this material is trademarked as "Polar Jade" and is of a translucent and rich green color seen in very few other specimens of nephrite. Large scale mining began there in 1995.

Gemological properties: Calcium magnesium iron silicate; Hardness: 6~6.5; Toughness: Exceptional; RI: 1.61~1.63; Density: 2.96; Polish Luster: greasy to vitreous; Fluorescence: none; Fracture: rarely seen.

Unit 16 Jewelry Commerce

Part 1 Dialogue

本章音频

Mary：What can I do for you, Madam?

Kate：Well, I hope you can help me. There is something wrong with this diamond ring, would you take a look?

Mary：Oh, dear. The diamond just comes off its location. Did you drop it or anything?

Kate：No. I bought it from your shop last week. And I hadn't worn it all the time. But I found the problem this morning.

Mary：Do you have proof of purchase?

Kate：The receipt and the guarantee? Here you are.

Mary：I'm terribly sorry. I apologize for the inconvenience. Now you need fill this form.

Kate：Next?

Mary：We'll have to send it to our factory to fix the diamond again.

Kate：How long will it be back? I do need it very soon.

Mary：It will be back in one week.

Kate：I was hoping you would be able to fix it in good time.

Mary：Very sorry. But we will treat as soon as possible.

Kate: Yes, thank you.
Mary: Once again. We're very sorry about this.
Kate: Well, I wait your word. Goodbye.
Mary: Goodbye.

Part 2 Fill the Blanks Basing on Your Gemological Knowledge

The words relating to jewelry commerce:

appraise:_____   estimate:_____   evaluate:_____

gem certificate:_____   invaluable:_____

valueless:_____   appreciate:_____   brand name:_____

trademark:_____   price tag:_____   salable ornament:_____

Part 3 Link the Paragraphs and the Relevant Pictures

(1) This is one of some absolutely spectacular labradorite pendants picked up in Quartzite, AZ. These are very flashy and the blue is out of this world. Labradorite is known to raise your level of consciousness and to connect one with the universal energy. It is also helpful in warding off negative energy. Location: Madagascar; Size: 2.25cm × 1.50cm; Weight: 0.1 lbs; Price: $15.00.

(2) Citrine, Weight: 11.34 ct; Size: 15.78mm × 11.96mm × 8.41mm; Color: Yellow golden; Shape: Octagon facet; Clarity: VVS; Treatment: None; Origin: Brazil; Gem ID: 241405.

(3) Ruby; Weight: 0.81 carat; Size: 5.5mm × 4.6mm Oval; Price: $445.00 This oval faceted ruby is a very nice medium red Burmese color (100% natural color). This gem has

not been irradiated and shows no indications of heat treatment. This brilliant ruby gemstone was faceted from a natural ruby crystal and has very faint natural inclusions which are visible to the unaided eye only under very close inspection.

(4) Star diopside; Weight: 33.83 ct; Size: 23.21mm×14.11mm×9.95mm; Color: Black; Shape: Oval cabochon; Opaque; Treatment: None; Origin: India. Diopside is a calcium magnesium silicate with a hardness of 5~6 on the Mohs' scale. Diopside is best known for the vivid green chrome diopside, but the black diopside exhibiting asterism or the star effect is also important. Diopside deposits are found in Burma, India, Madagascar, Sri Lanka, South Africa and the U.S.A.

(5) Sugilite; Weight: 32.23 ct; Size: 33.82mm×20.35mm×5.45mm; Color: Multicolor; Shape: Oval cabochon; Opaque; Treatment: None; Origin: Australia. Sugilite is an obscure and quite rare mineral named after the Japanese geologist, Ken-ichi Sugi, who discovered it in 1944. It is rarely found as crystals, usually being massive in form. It has a distinctive purple color and is usually opaque to translucent. Sugilite has sometimes been called purple turquoise although there is no connection between these minerals. Sugilite deposits are found in Australia, India, Japan, Canada and South Africa.

(6) Chrysoberyl Cat's Eye; Weight: 0.96 ct; Size: 6.06mm×4.69mm×3.53mm; Color: Green; Shape: Oval cabochon; Opaque; Treatment: None; Origin: India.

The most famous and valuable cat's eye gemstone is chrysoberyl cat's eye. In fact when the term cat's eye is used alone in the gem trade, it always refers to chrysoberyl cat's eye. All other types of cat's eye gems require an additional varieties designation, such as cat's eye apatite. Chrysoberyl cat's eye is so highly regarded due to its excellent hardness (8.5 on the Mohs' scale), superb luster and remarkably sharp cat's eye effect.

Part 4 Useful Phrases

ultraviolet phosphorescence   紫外磷光
ultraviolet fluorescence   紫外荧光
alexandrite stud earrings   变石(亚历山大石)耳钉
play of color   变彩效应
cleavage, fracture, parting of mineral   矿物的解理、断口、裂理
high pressure and high temperature treatment (HPHT)   高温高压处理
silver ornament   白银饰品
plated precious metal   镀色贵金属
characteristic inclusions   特征包裹体
specialized handtools and equipments   专业的工具和设备
crystal habit and surface features   晶体习性和表面特征
petrified (fossilized) wood   硅化木
cellular phone chain with iolite   带堇青石的手机链
colored glaze   琉璃

1.30 carat oval opal set in 14k yellow gold 镶嵌在14K黄金上的1.30克拉椭圆形欧泊

kunzite pendant 紫锂辉石吊坠

Part 5 Some Sentences

(1) Birthstones are gemstones given as a gift that symbolize the birth month of the receiver. The traditional birthstone of November is topaz, which is said to have benefits for its wearer.

(2) Assess the presentation of the stone. Examine clasps on necklaces and bracelets. Check settings on rings and earrings. If the metal is cheap and flimsy, the stone is probably not valuable.

(3) Modern jewelry of good quality is set in platinum or gold that is at minimum 14 karats. Some valuable gemstones are set in sterling silver, but this is very rare and usually reserved for semi-precious stones.

(4) Most gemstones are priced according to weight. The exceptions are some carvings and cabochons, which may be sold by the piece rather than by the carat, since the work required to produce them exceeds the cost of the material. But for the vast majority of gemstones, the price is computed on a per carat basis.

(5) Many people don't know that sapphires can be colorless, diamonds can be yellow, garnets can be purple and topaz can be pink. Color is not the most salient factor in the classification of gemstones, but it can affect the value of the stone. Purple or green garnets, red or pink topaz and red diamonds are the most rare and valuable color variations. In any stone, the most valuable will have a vivid and pure color.

Part 6 Short Paragraph

Gemstone Certificate

GIA (Gemological Institute of America) Diamond Reports/Dossier is the most used in the United States and many countries around the world. A report can clarify the authenticity of the stone. The report perfectly describes the evaluation of each of the key factors that influence the quality, beauty and value.

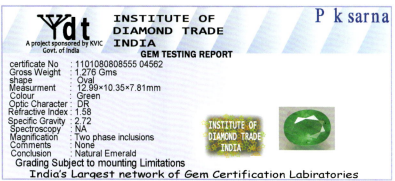

GIA DIAMOND DOSSIER

January 31, 2006

Laser Inscription Registry	GIA 14519089
Shape and Cutting Style	Round Brilliant
Measurements	4.35–4.38×1.61mm

GRADING RESULTS—GIA 4CS

Carat Weight	0.30 carat
Color Grade	G
Clarity Grade	IF
Cut Grade	Very Good

ADDITIONAL GRADING INFORMATION

Clarity Characteristics	
Finish	Natural
Polish	Very Good
Symmetry	Good
Fluorescence	None

GIA REPORT 14519089

The World's Foremost Authority in Gemology™

51141
20006558
RUSSIAN DIAMOND
CLUB INC./SMOLENSK
DIA
580 Fifth Avenue
New York, Ny 1036
0436507902
JOB:43101902

Jan 31,2006
GIA Report:14519089
Round Brilliant
4.35–4.38×2.61mm
Carat Weight:0.30carat
Color: G
Clarity:IF
Cut: Very Good
Total Depth:59.8%
Table Size:61%
Crown Angle:34.0°
Crown Height:13.0%
Pavilion Angle:41.2°
Pavilion Depth:43.5%
Star Length:50%
Lower Half:80%
Girdle:THN to MED (2.9%)
Culet:None
Polish:Very good
Symmetry:Good
Fluorescence:None
Comments:None

346692502

This Report is not a guarantee, valuation or appraisal and contains only the characteristics of the diamond described herein after it has been graded, tested, examined and analyzed by the GIA Laboratory and/or has been inscribed using the techniques and equipment used by the GIA Laboratory at the time of the examination and/or inscription. The recipient of this Report may wish to consult a credentialed jeweler or gemologist about the information contained herein.

IMPORTANT LIMITATIONS ON BACK
©2006 GEMOLOGICAL INSTITUTE OF AMERICA, INC.

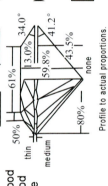

Profile to actual proportions.

In a general way, the gem certificate can insure the reality of the gem. The gem with certificate can be easily sold in the market. Now, we look at the following three certificates and learn the professional identification parameter.

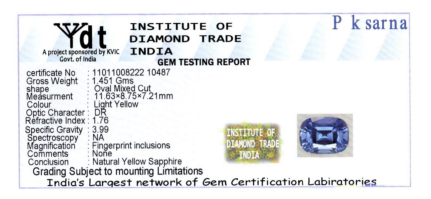

Part 7 Text

The Price of Gemstone

The price per carat of different gemstones can vary enormously, literally from $1 a carat to tens of thousands. Many factors influence the price per carat. Here is a concise summary of the 10 factors that determine gem prices:

Gem Variety

Some gemstone varieties—such as sapphire, ruby, emerald, garnet, tanzanite, spinel and alexandrite-command a premium price in the market, due to their superior gemstone characteristics and rarity. Other varieties, such as many types of quartz, are abundant in many locations around the world, and prices are much lower. But while the gem variety sets a general price range for a stone, the characteristics of the specific gem also have a major effect on the price per carat.

Color

In colored gemstones it is color that is the single most important determinant of value. Ideal colors vary by gem variety of course, but generally the colors that are most highly regarded are intense, vivid and pure. Gems that are too light or too dark are usually less desirable than those of medium tones. Thus a rich cornflower blue color in sapphire is more valuable than an inky blue-black or a pale blue.

Clarity

A gemstone that is perfectly clean, with no visible inclusions, will be priced higher. In general, the cleaner the stone, the better it's brilliance. So while it is true that the higher the clarity grade, the higher the value of the gem, inclusions that don't interfere with the brilliance and sparkle of a gem will not affect its value significantly. Note also that some gems, such as emerald, always have inclusions.

Cut and Polish

Gemstones should be cut with proper proportions to maximize the light that is returned to the eye. But gem cutters or lapidaries often have to make compromises when cutting a particular piece of material. If the gem color is quite light, cutting a deeper stone will provide a richer color. Conversely, a dark tone can be lightened by making a shallower cut. But in every case, the facets should meet cleanly and the surface should be well polished with no scratches.

Size

For some gemstone varieties, such as quartz, the price per carat is fairly constant as the weight of the stone increases. But in the case of many rarer gems, price does not increase in a linear way as the weight increases. Indeed for some gems, such as diamonds, the price per carat can increase exponentially as the carat size increases. According to this formula, a 1 carat stone may cost $1000 while a 2 carat stone may cost $4000. Though the formula is rarely so exact, good quality sapphires and rubies in larger sizes do tend to have a much higher per carat price.

Not only are larger stones more expensive, but gems cut in stock sizes—what are known in the trade as calibrated sizes—also tend to be more expensive. This is because more material has to be removed to achieve the calibrated size.

Shape

Some shapes tend to be priced higher than others, in part because of demand and in part because of material issues in cutting the specific shape. In general, round gems tend to command a premium in the market. Rounds are much less common than ovals, since ovals are usually cut to preserve as much weight of the raw material as possible. Cutting a round

gem normally requires a greater loss of the rough stone, and for very expensive materials like sapphire, ruby, alexandrite and others, this can have a significant effect on price.

Treatment

Gem treatments such as heating, fracture-filling, radiation and diffusion significantly improve the appearance of many gemstones, and these treatments are now routine for many commercial-grade stones. A treated stone is always less expensive than a similar untreated stone. But most of the stones that are routinely treated—such as ruby and sapphire—are now very rare in untreated form, and the untreated stones fetch a market price out of the reach of most consumers. If you prefer to buy an untreated stone, you do have many choices. A number of popular gems, such as tourmaline, spinel, amethyst and garnet are almost never treated.

Origin

Strictly speaking, a fine natural gem is a fine gem, regardless of its country or region of origin. The reality of the market is that certain gem varieties from locations such as Burma, Kashmir, Sri Lanka and Brazil, command a premium price in the market. It is difficult to say whether this premium is justified, especially with so many fine gems coming from Africa.

Fashion

Some gems, such as blue sapphire, are always in fashion. But some gems become fashionable for short periods, with resulting price increases. Recently we've seen andesine labradorite and diaspore in the spotlight, and strong interest in rutilated quartz. Some very fine gems, such as natural spinel, actually have lower than expected prices because limited supply means that the gems are not promoted heavily in the market.

Supply Chain

The gem trade is a business and everyone in the supply chain-from the mine to the jewelry retailer is trying to turn a profit. Gemstones can pass through many hands on the way from the mine to the consumer, and the more brokers and distributors that handle the product, the higher the final price will be. So in fact the same gemstone may carry a price that varies by as much as 200%, depending on who you buy it from.

Part 8 Words and Expressions

proof[pru:f]	n. 证据
inconvenience[,ɪnkən'vi:njəns]	n. 麻烦,不方便之处
appraise[ə'preɪz]	v. 评价

estimate['estɪmeɪt]	v. 估计,估价,评估　n. 估计 估价,评估
evaluate[ɪ'væljueɪt]	vt. 评价,估计,求…的值　v. 评价
certificate[sə'tɪfɪkɪt]	n. 证书,证明书　vt. 发给证明书,以证书形式授权给……
gem certificate	宝石鉴定证书
invaluable[ɪn'væljuəbl]	adj. 无价的,价值无法衡量的
valueless['vælju:ləs]	adj. 不足道的
appreciate[ə'pri:ʃɪeɪt]	vt. 赏识,鉴赏,感激　vi. 增值,涨价
brand name	n. 商标,品牌
trademark['treɪdmɑ:k]	n. 商标
price tag	n. 价格标签
salable['seɪləbl]	adj. 适于销售的
salable ornament	畅销饰品
absolutely['æbsəlu:tlɪ]	adv. 完全地,绝对地
labradorite['læbrə,dɔ:raɪt]	n. 拉长石,富拉玄武岩
quartzite['kwɔ:tsaɪt]	n. 石英岩,硅岩
consciousness['kɒnʃəsnəs]	n. 意识,知觉,自觉,觉悟,个人思想
diopside[daɪ'ɒpsaɪd]	n. 透辉石
star diopside	星光辉石
magnesium[mæg'ni:zɪəm]	n. 镁(元素符号 Mg)
calcium magnesium silicate	钙镁硅酸盐
asterism['æstərɪzəm]	n. 星光效应
sugilite	译音"苏纪石","苏纪莱石",矿物名为硅铁锂钠石
multicolor['mʌltɪ'kʌlə]	n. 多色
obscure[əb'skjʊə]	adj. 暗的,朦胧的,模糊的
varietal[və'raɪətl]	adj. 品种的,变种的
superb[sju:'pə:b]	adj. 庄重的,堂堂的,华丽的,极好的
presentation[,prezn'teɪʃn]	n. 介绍,陈述,赠送,表达
clasp[klɑ:sp]	n. 扣子,钩,紧握,抱住　v. 扣紧,紧握,搂抱,密切合作
flimsy['flɪmzɪ]	adj. 易坏的,脆弱的,浅薄的,没有价值的,不足信的,(人)浮夸的
sterling['stə:lɪŋ]	n. 英国货币,标准纯银 adj. 英币的,纯银制的,纯正的
sterling silver	n. 标准纯银(纯度为百分之 92.5)
carving['kɑ:vɪŋ]	n. 雕刻品,雕刻
salient['seɪljənt]	adj. 易见的,显著的,突出的,跳跃的 n. 凸角,突出部分
GIA(Gemological Institute of America)	美国宝石学院

dossier['dɒsieɪ]	n. 档案,卷宗
authenticity[ˌɔːθen'tɪsəti]	n. 确实性,真实性
influence['ɪnfluəns]	n. 影响,感化,势力,有影响的人(或事),(电磁)感应 vt. 影响
parameter[pə'ræmɪtə]	n. 参数,参量,起限定作用的因素
professional identification parameter	专业鉴定参数
enormously[ɪ'nɔːməsli]	adv. 非常地,巨大地
literally['lɪtərəli]	adv. 照字面意义,逐字地
concise[kən'saɪs]	adj. 简明的,简练的
command[kə'mɑːnd]	n. 命令,掌握,司令部 v. 命令,指挥,克制,支配,博得,俯临
premium['priːmɪəm]	n. 额外费用,奖金,奖赏,保险费,(货币兑现的)贴水
determinant[dɪ'təːmɪnənt]	adj. 决定性的
intense[ɪn'tens]	adj. 强烈的,剧烈的,热切的,热情的,激烈的
cornflower['kɔːnflaʊə]	n. [植]矢车菊
inky['ɪŋki]	adj. 漆黑的,墨水的,给墨水弄污的
clarity['klærɪti]	n. 清楚,透明
visible inclusion	可见包裹体
polish['pɒlɪʃ]	n. 磨光,光泽,上光剂,优雅,精良 vt. 擦亮,发亮,磨光,推敲 vi. 发亮,变光滑
compromise['kɒmprəmaɪz]	n. 妥协,折衷 v. 妥协,折衷,危及……的安全
conversely[kən'vəːsli]	adv. 倒地,逆地
exponentially[ˌekspə'nenʃəli]	n. 指数,演奏者,例子 adj. 指数的,幂数的
stock size	(鞋、帽、成衣等的)现货号码(尺寸)
calibrate['kælɪbreɪt]	v. 校准
loss[lɒs]	n. 损失,遗失,失败,输,浪费,错过,[军]伤亡,降低
roughstone	原石,原料
treatment['triːtmənt]	n. 处理,待遇,对待,治疗
fracture-filling	裂隙充填
commercial-grade stone	商业级宝石
fetch[fetʃ]	vt. 接来,取来,带来,售得,引出,吸引,到达,演绎出 vi. 取物,绕道而行 n. 取得,拿,诡计,魂
consumer[kən'sjuːmə]	n. 消费者
andesine['ændɪziːn]	n. 中长石
andesine labradorite	n. 中长拉长石
diaspore['daɪəspɔː]	n. 水铝石,传播体
spotlight['spɒtlaɪt]	n. 聚光灯
broker['brəʊkə]	n. 掮客,经纪人

Part 9 Check Your Understanding

Exercise 1 Answer the following questions

(1) How do we introduce the gemstone to buyer?

(2) Please talk about GIA and GIA Diamond Dossier.

(3) Which items were appeared on the gem certificate ordinarily?

(4) Which countries were the Sugilite produced?

Exercise 2 Fill in the blanks

(1) Please illustrate the factors that determine gem prices: _____.

(2) In colored gemstones it is _____ that is the most important determinant of value.

(3) Gemstones should be cut with _____ to maximize the light that is returned to the eye.

(4) Gem treatments such as _____, _____, _____ and _____ significantly improve the appearance of many gemstones.

Exercise 3 Translation

(1) This brilliant ruby gemstone was faceted from a natural ruby crystal and has very faint natural inclusions which are visible to the unaided eye only under very close inspection.

(2) The price per carat of different gemstones can vary enormously, literally from $1 a carat to tens of thousands. Many factors influence the price per carat.

(3) Strictly speaking, a fine natural gem is a fine gem, regardless of its country or region of origin. The reality of the market is that certain gem varieties from locations such as Burma, Kashmir, Sri Lanka and Brazil, command a premium price in the market.

(4) 如果宝石带有鉴定证书,顾客选购时会比较放心,不用担心买到假货。

(5) 影响宝石价格的因素较多,如宝石品种、颜色、净度、切工、个人爱好等。

(6) 珠宝文化及寓意对珠宝市场有较大的影响。

Part 10 Self-Study Material

Birthstones

January Garnet

Garnet, the birthstone for January, signifies friendship and trust and is the perfect gift for a friend. Garnet, derived from the word granatum, means seed, and is called so because of the gemstone's resemblance to a pomegranate seed. References to the gemstone dates back to 3100 B.C., when the Egyptians used garnets as inlays jewelry. Today, the most important sources for garnet are Africa, Sri Lanka, and India.

February Amethyst

Amethyst is said to keep the wearer clear-headed and quick-witted. Throughout history, the gemstone has been associated with many myths, legends, religions, and numerous cultures. English regalia were even decorated with amethysts during the Middle Ages to symbolize royalty. Amethyst is purple quartz, a beautiful blend of violet and red that can found in every corner of the earth. Today, Brazil is the primary source of this gemstone, fine material can be found elsewhere, especially in Zambia.

March Aquamarine

The name of aquamarine is derived from the Latin word aqua, meaning water, and sea. This gemstone was believed to protect sailors, as well as to guarantee a safe voyage. The serene color of aquamarine is said to cool the temper, allowing the wearer to remain calm and levelheaded. This gemstone is mined mainly in Brazil, but also is found in Nigeria, Madagascar, Zambia, Pakistan, and Mozambique.

April Diamond

As the April birthstone, diamonds are the ideal gift for a loved one. And now you have more choices than ever. Get creative and give the ultimate gift of beauty: a fancy-color diamond. Fancy-color diamonds are natural, rare and truly exotic gem of the earth. Diamonds in hues of yellow, red, pink, blue, and green range in intensity from faint to vivid and generally the more saturated the color, the higher the value.

May Emerald

As the birthstone for May, the emerald, a symbol of rebirth, is believed to grant the owner foresight, good fortune, and youth. Emerald, derived from the word smaragdus, meaning greenin Greek, was mined in Egypt as early as 330 B. C. Today, most of the world's emeralds are mined in Colombia, Brazil, Afghanistan, and Zambia. The availability of high-quality emerald is limited; consequently, treatments to improve clarity are performed regularly.

June Pearl

Historically, pearls have been used as an adornment for centuries. They were one of the favorite gem materials of the Roman Empire; later in Tudor England, the 1500s were known as the pearl age. Pearls are unique as they are the only gems from living sea creatures and require no faceting or polishing to reveal their natural beauty. In the early 1900s, the first successful commercial culturing of round saltwater pearls began. Since the 1920s, cultured pearls have almost completely replaced natural pearls in the market.

July Ruby

There's no better way to demonstrate your love than by giving a ruby in celebration of a July birthday. Rubies arouse the senses, stir the imagination, and are said to guarantee health, wisdom, wealth and success in love. Ruby is a variety of the gems species corundum. It is harder than any natural gemstone except diamond, which means a ruby is durable enough for everyday wear. Fine-quality ruby is extremely rare, and the color of the gem is most important to its value. The most prized color is a medium or medium dark vivid red or slightly purplish red. If the gem is too light or has too much purple or orange, it will be called a fancy-color sapphire.

August Peridot

Peridot is said to host magical powers and healing properties to protect against nightmares and to bring the wearer power, influence, and a wonderful year. As peridot is a gemstone that forms deep inside the Earth and brought to the surface by volcanoes, in Hawaii, peridot symbolizes the tears of Pele, the goddess of fire and volcanoes. Today, most of the peridot supply comes from Arizona; other sources are China, Myanmar, and Pakistan.

September Sapphire

Sapphire, the September birthstone, has been popular since the Middle Ages and, according to folklore, will protect your loved ones from envy and harm. Medieval clergy wore sapphires to symbolize heaven, while commoners thought the gem attracted heavenly blessings. The most prized colors are a medium to medium dark blue or slightly violet-blue. Sapphire is a variety of the gem species corundum and occurs in all colors of the rainbow. Pink, purple, green, orange, or yellow corundum are known by their color (pink sapphire, green sapphire).

October Opal

The name of opal derives from the Greek Opallos, meaning "to see a change (of color)". Opals range in color from milky white to black with flashes of yellow, orange, green, red, and blue. An opal's beauty is the product of contrast between its color play and its background. Opal is a formation of non-crystalline silica gel that seeped into crevices in the sedimentary strata.

November Topaz

Topaz is a gemstone available in a rich rainbow of colors. Prized for several thousand years in antiquity, all yellow gems in antiquity were called topaz. Often confused with citrine quartz (yellow) and smoky quartz (brown), quartz and topaz are separate and unrelated mineral species. Topaz also comes in yellow, pink, purple, orange, and the many popular blue tones.

December Turquoise

Turquoise, also named Turkish stone, originated in the thirteenth century and describes one of the oldest known gemstones. Turquoise varies in color from greenish blue to sky blue shades and its transparency ranges from translucent to opaque. Turquoise is plentiful and is available in a wide range of sizes. It is most often used for beads, cabochons, carvings, and inlays. Although its popularity fluctuates in fashion, it is a perennial favorite in the American Southwest.

附录1 珠宝玉石英文单词表

A

actinolite[ˌæk'tɪnəlaɪt]	n. 阳起石
agate['æget]	n. 玛瑙
albite['ælbɪt]	n. 钠长石
alexandrite [ˌælɪg'zɑːndraɪt]	n. 变石
alexandrite cat's-eye	变石猫眼
almandite [ˌælmən'daɪt]	n. 铁铝榴石,贵榴石
amazonite [ˌæmə'zəʊnaɪt]	n. 天河石
amber ['æmbə]	n. 琥珀
amblygonite ['æmblɪgənaɪt]	n. 磷铝锂石
amethyst ['æməθɪst]	n. 紫晶
andalusite ['ændəluːsaɪt]	n. 红柱石
andradite ['ændrədaɪt]	n. 钙铁榴石
apatite ['æpətaɪt]	n. 磷灰石
apophyllite[ə'pɒfɪlaɪt]	n. 鱼眼石
aquamarine [ˌækwəmə'riːn]	n. 海蓝宝石
augite['ɔːdʒaɪt]	n. 普通辉石
aventurine quartz	东陵石
aventurine [æ'ventʃərɪn]	n. 砂金石
axinite ['æksɪnaɪt]	n. 斧石

B

banded agate	条纹玛瑙
barite ['beəraɪt]	n. 重晶石
benitoite[bɪnɪ'tɔɪt]	n. 蓝锥矿
beryl ['berəl]	n. 绿柱石
black opal	n. 黑欧泊
brazilianite [brə'zɪlɪənaɪt]	n. 磷铝钠石

C

calamine ['kæləmaɪn]	n. 菱锌矿

carnelian [kɑːˈniːliən]	n. 红玉髓
cassiterite [kəˈsɪtəraɪt]	n. 锡石
celestite [ˈselɪstaɪt]	n. 天青石
chalcedony [kælˈsedənɪ]	n. 玉髓
charoite [ˈtʃɑːrəait]	n. 查罗石, 紫硅碱钙石
chiastolite [kaɪˈæstəˌlaɪt]	n. 空晶石
chrysoberyl [ˈkrɪsəberɪl]	n. 金绿宝石
chrysoberyl cat's-eye	猫眼
chrysocolla [krɪsəˈkɒlə]	n. 硅孔雀石
chryopase [ˈkrɪsəpreɪz]	n. 绿玉髓, 澳洲玉
citrine [ˈsɪtrɪn]	n. 黄晶
coral [ˈkɒrəl]	n. 珊瑚
cordierite [ˈkɔːdɪəˌraɪt]	n. 堇青石
corundum [kəˈrʌndəm]	n. 刚玉, 金刚砂
crocidolite [krəʊˈsɪdəˌlaɪt]	n. 青石棉
crystal [ˈkrɪstl]	n. 水晶
cultured pearl	养殖珍珠

D

danburite [ˈdænbəraɪt]	n. 赛黄晶
datolite [ˈdætəlaɪt]	n. 硅硼钙石
demantoid [dɪˈmæntɔɪd]	n. 翠榴石
diamond [ˈdaɪəmənd]	n. 钻石
diopside [daɪˈɒpsaɪd]	n. 透辉石
dioptase [daɪˈɒpteɪs]	n. 透视石
dolomite [ˈdɒləmaɪt]	n. 白云石
dushan-jade	n. 独山玉

E

emerald [ˈemərəld]	n. 祖母绿
enstatite [ˈenstətaɪt]	n. 顽火辉石
epidote [ˈepɪdəʊt]	n. 绿帘石
euclase [ˈjuːkleɪs]	n. 蓝柱石

F

fabulite-artificial product	人造钛酸锶
feldspar [ˈfeldspɑː]	n. 长石
fire agate	火玛瑙
fire opal	火欧泊

fluorite[ˈfluərait]	n. 萤石
freshwater cultured pearl	淡水养殖珍珠
freshwater pearl	淡水珍珠

G

garnet[ˈgɑːnɪt]	n. 石榴石
GGG＝Gadolinium Gallium garnet	人造钆镓榴石
glass artificial product	玻璃
goethite[ˈgəʊθaɪt]	n. 乌钢石，针铁矿
green quartz	绿水晶
grossularite[ˈgrɒsjʊlərait]	n. 钙铝榴石

H

hawk's-eye	鹰眼石
hematite [ˈhemətait]	n. 赤铁矿
hessonite[ˈhesəˌnait]	n. 钙铝榴石，肉桂石
howlite[ˈhaʊlait]	n. 羟硅硼钙石
hydrogrossular[ˌhaɪdrəˈgrɒsjʊlə]	n. 水钙铝榴石

I

idocrase[ˈɪdəʊˌkreɪs]	n. 符山石
iolite [ˈaɪəlait]	n. 堇青石
ivory [ˈaɪvəri]	n. 象牙

J

jadeite[ˈdʒeɪdait]	n. 翡翠
jasper [ˈdʒespə(r)]	n. 碧玉
jet [dʒet]	n. 煤精

K

kornerupine [ˌkɔːnəˈruːpin]	n. 柱晶石
kyanite[ˈkaɪəˌnait]	n. 蓝晶石

L

labradorite[ˌlæbrəˈdɒrait]	n. 拉长石
lapis lazuli [ˈlæpɪsˈlæzjʊlai]	n. 青金石
lazulite [ˈlæzjʊˌlait]	n. 天蓝石

M

malachite[ˈmæləkaɪt]	n. 孔雀石
marble[ˈmɑːbl]	n. 大理石
maw-sit-sit	n. 钠长硬玉,莫西西
melanite[ˈmeləˌnaɪt]	n. 黑榴石
moldavite [ˈməʊldəvaɪt]	n. 玻璃陨石
moonstone[ˈmuːnstəʊn]	n. 月光石
moss agate	苔纹玛瑙
mother of pearl	n. 贝壳

N

natural glass	天然玻璃
nephrite [ˈnefraɪt]	n. 软玉,和田玉

O

obsidian[ɑbˈsɪdiən]	n. 黑曜岩
oligoclase[ˈɒlɪgəʊkleɪs]	n. 奥长石
onyx[ˈɒnɪks]	n. 缟玛瑙
opal [ˈəʊpl]	n. 欧泊,蛋白石
orthoclase [ˈɔːθəkleɪs]	n. 正长石

P

pearl [pɜːl]	n. 珍珠
peridot [ˈperidɒt]	n. 橄榄石
petrified wood	木化石,硅化木
phenakite [ˈfenəkaɪt]	n. 硅铍石
plagioclase [ˈpleɪdʒɪəkleɪs]	n. 斜长石
plastic [ˈplæstɪk]	n. 塑料
prase [preiz]	n. 绿石英
prehnite [ˈpreɪnaɪt]	n. 葡萄石
pyrope[ˈpaɪrəʊp]	n. 镁铝榴石

Q

quartz[kwɔːts]	n. 石英
quartzite[ˈkwɔːtsait]	n. 石英岩,密玉

R

rhodolite[ˈrəʊdəlaɪt]	n. 铁镁铝榴石

rhodonite[ˈrəʊdənaɪt]	n. 蔷薇辉石
rose quartz	芙蓉石
ruby [ˈruːbi]	n. 红宝石

S

sapphire[ˈsæfaɪə]	n. 蓝宝石
sardonyx [ˈsɑːdənɪks]	n. 缠丝玛瑙,红条纹玛瑙
scapolite [ˈskæpəlaɪt]	n. 方柱石
seawater cultured pearl	海水养殖珍珠
seawater pearl	海水珍珠
serpentine [ˈsɜːpəntaɪn]	n. 蛇纹石,岫玉
silicified wood	硅化木
sillimanite [ˈsɪlɪmənaɪt]	n. 矽线石
sinhalite [ˈsɪnhəlaɪt]	n. 硼铝镁石
smithsonite[ˈsmɪθsənaɪt]	n. 菱锌矿
smoky quartz	烟晶
sodalite [ˈsəʊdəlaɪt]	n. 方钠石
spessartite [ˈspesətaɪt]	n. 锰铝榴石
sphene[sfiːn]	n. 楣石
spinel [ˈspɪnəl]	n. 尖晶石
spodumene [ˈspɒdjʊmiːn]	n. 锂辉石
sunstone[ˈsʌnstəʊn]	n. 日光石
synthetic alexandrite	合成变石
synthetic amethyst	合成紫晶
synthetic chrysoberyl	合成金绿宝石
synthetic critrine	合成黄晶
synthetic crystal	合成水晶
sythetic cubic zircon	合成立方氧化结石
synthetic diamond	合成钻石
synthetic emerald	合成祖母绿
synthetic green quartz	合成绿水晶
synthetic opal	合成欧泊
synthetic ruby	合成红宝石
synthetic rutile	合成金红石
synthetic sapphire	合成蓝宝石
synthetic somky quartz	合成烟晶
synthetic spinel	合成尖晶石
synthetic turquoise	合成绿松石

T

taaffeite [ˈtɑːfaɪt]	n. 塔菲石
talc [tælk]	n. 滑石
tanzanite [ˈtænzənaɪt]	n. 坦桑石
tiger's-eye	虎睛石
topaz [ˈtəʊpæz]	n. 托帕石
tortoiseshell [ˈtɔːtəsʃel]	n. 龟甲,玳瑁
tourmaline [ˈtʊəməlɪn]	n. 碧玺,电气石
tremolite [ˈtreməlaɪt]	n. 透闪石
turquoise [ˈtɜːkwɔɪz]	n. 绿松石

U

uvarovite [uːˈvɑːrəvaɪt]	n. 钙铬榴石

Y

YAG＝Yttrium Aluminum Garnet	人造钇铝榴石

Z

zircon [ˈzɜːkɒn]	n. 锆石
zoisite [ˈzɔɪsaɪt]	n. 黝帘石

附录2 化学元素(部分)中英文对照表

序号	符号	中文	英文	序号	符号	中文	英文
1	H	氢	Hydrogen[ˈhaɪdrədʒən]	23	V	钒	Vanadium[vəˈneɪdɪəm]
2	He	氦	Helium[ˈhiːlɪəm]	24	Cr	铬	Chromium[ˈkrəʊmɪəm]
3	Li	锂	Lithium[ˈlɪθɪəm]	25	Mn	锰	Manganese[ˈmæŋɡəniːz]
4	Be	铍	Beryllium[bəˈrɪlɪəm]	26	Fe	铁	Iron[ˈaɪən]
5	B	硼	Boron[ˈbɔːrɒn]	27	Co	钴	Cobalt[ˈkəʊbɔːlt]
6	C	碳	Carbon[ˈkɑːbən]	28	Ni	镍	Nickel[ˈnɪkl]
7	N	氮	Nitrogen[ˈnaɪtrədʒən]	29	Cu	铜	Copper[ˈkɒpə(r)]
8	O	氧	Oxygen[ˈɒksɪdʒən]	30	Zn	锌	Zinc[zɪŋk]
9	F	氟	Fluorine[ˈflɔːriːn]	31	Ga	镓	Gallium[ˈɡælɪəm]
10	Ne	氖	Neon[ˈniːɒn]	32	Ge	锗	Germanium[dʒɜːˈmeɪnɪəm]
11	Na	钠	Sodium[ˈsəʊdɪəm]	33	As	砷	Arsenic[ˈɑːsnɪk]
12	Mg	镁	Magnesium[mæɡˈniːzɪəm]	34	Se	硒	Selenium[səˈliːnɪəm]
13	Al	铝	Aluminum[əˈljuːmɪnəm]	35	Br	溴	Bromine[ˈbrəʊmiːn]
14	Si	硅	Silicon[ˈsɪlɪkən]	36	Kr	氪	Krypton[ˈkrɪptɒn]
15	P	磷	Phosphorus[ˈfɒsfərəs]	37	Rb	铷	Rubidium[ruːˈbɪdɪəm]
16	S	硫	Sulfur[ˈsʌlfə]	38	Sr	锶	Strontium[ˈstrɒntɪəm]
17	Cl	氯	Chlorine[ˈklɔːriːn]	39	Y	钇	Yttrium[ˈɪtrɪəm]
18	Ar	氩	Argon[ˈɑːɡɒn]	40	Zr	锆	Zirconium[zɜːˈkəʊnɪəm]
19	K	钾	Potassium[pəˈtæsɪəm]	41	Nb	铌	Niobium[naɪˈəʊbɪəm]
20	Ca	钙	Calcium[ˈkælsɪəm]	42	Mo	钼	Molybdenum[məˈlɪbdənəm]
21	Sc	钪	Scandium[ˈskændɪəm]	43	Tc	锝	Technetium[tekˈniːʃɪəm]
22	Ti	钛	Titanium[tɪˈteɪnɪəm]	44	Ru	钌	Ruthenium[ruːˈθiːnɪəm]

续上表

序号	符号	中文	英文	序号	符号	中文	英文
45	Rh	铑	Rhodium['rəudiəm]	73	Ta	钽	Tantalum['tæntələm]
46	Pd	钯	Palladium[pə'leidiəm]	74	W	钨	Tungsten['tʌŋstən]
47	Ag	银	Silver['sɪlvə]	75	Re	铼	Rhenium['riːniəm]
48	Cd	镉	Cadmium['kædmiəm]	76	Os	锇	Osmium['ɒzmiəm]
49	In	铟	Indium['ɪndiəm]	77	Ir	铱	Iridium[ɪ'rɪdiəm]
50	Sn	锡	Tin[tin]	78	Pt	铂	Platinum['plætɪnəm]
51	Sb	锑	Antimony['æntɪməni]	79	Au	金	Gold[gəuld]
52	Te	碲	Tellurium[te'ljuəriəm]	80	Hg	汞	Mercury['mɜːkjəri]
53	I	碘	Iodine['aɪədiːn]	81	Tl	铊	Thallium['θæliəm]
54	Xe	氙	Xenon['zenɒn]	82	Pb	铅	Lead[liːd]
55	Cs	铯	Cesium['siːzɪəm]	83	Bi	铋	Bismuth['bɪzməθ]
56	Ba	钡	Barium['beəriəm]	84	Po	钋	Polonium[pə'ləuniəm]
57	La	镧	Lanthanum['lænθənəm]	85	At	砹	Astatine['æstətiːn]
58	Ce	铈	Cerium['sɪəriəm]	86	Rn	氡	Radon['reɪdɒn]
59	Pr	镨	Praseodymium[,preɪzɪəu'dɪmiəm]	87	Fr	钫	Francium['frænsiəm]
60	Nd	钕	Neodymium[,niːəu'dɪmiəm]	88	Ra	镭	Radium['reɪdiəm]
61	Pm	钷	Promethium[prə'miːθiəm]	89	Ac	锕	Actinium[æk'tɪniəm]
62	Sm	钐	Samarium[sə'meəriəm]	90	Th	钍	Thorium['θɔːriəm]
63	Eu	铕	Europium[juə'rəupiəm]	91	Pa	镤	Protactinium[,prəutæk'tɪniəm]
64	Gd	钆	Gadolinium[,gædə'lɪniəm]	92	U	铀	Uranium[ju'reɪniəm]
65	Tb	铽	Terbium['tɜːbiəm]	93	Np	镎	Neptunium[nep'tjuːniəm]
66	Dy	镝	Dysprosium[dɪs'prəuziəm]	94	Pu	钚	Plutonium[pluː'təuniəm]
67	Ho	钬	Holmium['həulmiəm]	95	Am	镅	Americium[,æmə'rɪsiəm]
68	Er	铒	Erbium['ɜːbiəm]	96	Cm	锔	Curium['kjuəriəm]
69	Tm	铥	Thulium['θuːliəm]	97	Bk	锫	Berkelium[bɜː'kiːliəm]
70	Yb	镱	Ytterbium[ɪ'tɜːbiəm]	98	Cf	锎	Californium[,kælɪ'fɔːniəm]
71	Lu	镥	Lutetium[luː'tiːʃiəm]	99	Es	锿	Einsteinium[aɪn'staɪniəm]
72	Hf	铪	Hafnium['hæfniəm]	100	Fm	镄	Fermium['fɜːmiəm]

参 考 译 文

Unit 1　Gems and Gemstones Instruction
（宝石和宝石材料介绍）

Part 5　Some Sentences（一些句子）

（1）宝石学是一门研究宝石材料的性质、产地和成因的科学。珠宝商业中，像珠宝商、珠宝消费者、首饰设计师和评估师，还有古董商人、拍卖行的编目人员等通常需要了解宝石学知识，从事这些工作的人都需要能鉴定宝石和宝石材料，并能描述其性质。

（2）采用水热法在桂林合成了红宝石晶体，对晶体的生长设备结合工艺条件进行了研究，对晶体的宝石学特征进行了测定。

（3）这篇文章介绍了刻面水晶的抛光工艺和钻孔工艺，并对加工设备和各种抛光材料的性能、特点作了分析。

Part 6　Short Paragraph（短文）

Important Qualities of a Gemstone（宝石材料的重要特性）

一种宝石材料通常因其具有传统的"三大基本特征"——美丽、耐久、稀少而被人们珍视。另外，宝石的价值还取决于社会对宝石的接受程度和宝石的便携性。

美丽是宝石材料与可见光相互作用的结果。宝石的美丽程度因宝石材料不同而异，尤其受宝石透明度的影响。

耐久性与结构有关，取决于宝石材料抵抗刻划能力的硬度、抵抗敲击发生破裂能力的韧性和抗高温、压力、撞击及化学作用等外界因素而具有的稳定性。

稀少是指宝石材料可获得的难易程度，评估其稀少性要考虑市场经济中的供给和需求。

Part 7　Text（课文）

Gems and Gemstones Instruction（宝石和宝石材料的介绍）

什么是宝石材料？什么是宝石？

你可曾见过钻石戒指？欧泊项链？珍珠耳环？钻石、欧泊、珍珠都是不同种类的宝石材料。宝石材料是一种用来制作珠宝首饰、装饰品或艺术品的矿物、岩石或有机材料。宝石是一种经过切割、抛光后展露美丽的宝石材料，例如钻石。欧泊和珍珠可能没有经过切割或抛光就用来做珠宝首饰或艺术品，那它们就是宝石材料而不是宝石。

这个吊坠由已切割并抛光的数颗钻石和一颗欧泊组成。你认为这些钻石是宝石还是宝石材料？欧泊呢？答案是：钻石是宝石，欧泊是宝石材料。

宝石是天然具有美丽、耐久、稀少等性质的矿物或有机材料。

天然的意思就是这些材料不是人工创造的或是人为辅助形成的。联邦贸易委员会（FTC）规定：如果是非天然宝石材料，在广告和市场上一般都必须有"实验室生长""合成""养殖""人造"等修饰词。人造"宝石"与其模仿的天然材料有相同的化学性质、光学性质和物理性质，但是它们的稀有性和价值大不相同。通过以上任何修饰词都可以确定有疑问的材料不是天然的。

矿物是由原子在三维空间有规律排列组成的而有特定化学式的结晶固体。

在哪里找宝石？

宝石材料在世界各地都有产出。钻石可出现在地球深部的金伯利岩中。电气石和绿柱石的围岩被剥蚀后，可以在河床中找到它们。石榴石常出现在高温下形成的片麻岩中。

电气石常出现多种颜色，常常在一个晶体上出现多种颜色。电气石晶体与石英共生时出现三彩，三彩的意思就是由三种颜色组成。

宝石是怎么形成的？

宝石材料是在一些特定的不同方式下形成的。它们的颜色通常是由形成时的化学成分决定的。

绿松石是由含铜、铝、磷的矿石在水的淋滤作用下形成的。绿松石通常出现在干燥的或沙漠环境中，如美国的西南部。美国西南印第安人的美丽首饰可能广为人知，那些首饰通常是绿松石做的。

青金石是一种岩石而不是矿物。当岩浆从地下侵入岩石形成青金石时，炽热的岩浆使围岩熔融再凝固，这个过程中就形成了由青金石、黄铁矿和方解石组成的新的深蓝色青金岩。

人们通常认为石榴石是红色的，但实际上石榴石可以出现从黄色到黑色的一系列颜色。变色石榴石在日光灯和白炽灯下分别显示不同的颜色。

石榴石通常是在高温高压变质岩中形成的。石榴石可以呈现各种颜色。它们的颜色是由熔融矿物的化学成分决定的。红色石榴石，即镁铝榴石，它的颜色是它形成时候的熔融化学混合物——硅酸镁的颜色。

玉石是一种珍贵的材料，常用在丧葬、王冠、首饰中，或被有象形文字的文化珍视，如中美洲的奥梅克人和中国人。

玉是一种宝石，常用翡翠和软玉作为原料来切割抛磨。软玉和翡翠中的主要矿物都存在于地球深部的变质岩中。

合成宝石

在实验室中可以制造宝石材料和宝石。科学家们在实验室里尽力模仿宝石材料和宝石的自然生长条件，以使仿制品与天然品有相似的性质。在人工环境下可以生产绿松石、蓝宝石和红宝石。

有一种方法可以生长红宝石，即将带有籽晶的籽晶杆下降到熔融矿液中然后提回，一遍一遍地重复此操作，籽晶杆末端将长成大粒晶体，从籽晶杆上分离晶体、切割并抛光将得到红宝石。

宝石分类

科学家和宝石学家已经提出很多种分类宝石的方法：珍贵宝石或次珍贵宝石、天然宝石或合成宝石以及有机宝石或无机宝石。

珍贵宝石，如几个世纪以来，霍普钻石几易货主，不同的主人多次切割、抛光，使其呈现现在的样子。

珍贵宝石或次珍贵宝石

这种分类是基于宝石的美丽、耐久和稀少性。它主要是针对宝石,而不是宝石材料。珍贵宝石一般都是最美丽、最耐久和最稀少的。珍贵宝石包括钻石、红宝石、蓝宝石、祖母绿、海蓝宝石、托帕石和欧泊。而次珍贵宝石相对来说没有那么美丽、耐久和稀少。它们的摩氏硬度一般低于8,也就是说它们比较容易被刻划。绿松石、玉石、青金石和琥珀都是次珍贵宝石。

一个蓝宝石矿主手捧几个细小的宝石材料,这些都是她在马达加斯加的一个矿中找到的。

是不是就认为托帕石比绿松石要漂亮呢?很难说。将宝石分为珍贵或次珍贵难以得到大家的认同,因为对美丽总是有不一致的观点。

天然宝石或合成宝石

天然宝石是在地球上形成的,而合成宝石是在实验室形成的。有些宝石是两种环境都有。祖母绿、石榴石、红宝石、蓝宝石和钻石在地球和实验室中都可以形成。

你见过含有昆虫的琥珀或者是含有植物的琥珀吗?这个图片显示了早期琥珀的形成和蚂蚁多容易粘在黏稠的树脂中。

有机宝石或无机宝石

有机宝石是一种由生物形成的宝石。珍珠是由贝类或者蚌类动物生成。琥珀是由树的松油演化来的。珊瑚是由海洋中大量的生物组成。珍珠、琥珀和珊瑚都是有机宝石材料。无机宝石包括钻石、蓝宝石和红宝石,它们都由矿物组成,没有生物作用。

宝石文化

(1)珍珠文化:几百年以前,沿着美国东海岸,印第安人用珍珠做首饰,男人和女人都佩戴珍珠首饰,包括珍珠耳坠。据传说,波卡洪塔斯将收藏的珍珠作为礼物送给她的父亲波瓦坦。

(2)绿松石文化:波斯开采绿松石已有几千年的历史了。它出口到其他地方,如埃及,那里的法老用绿松石。后来,在美国西南部也发现了绿松石。现在美国是最大的绿松石供应地。

(3)石英文化:石英是地壳中第二常见的矿物。石英有很多种类,如粉晶、烟晶和紫晶。石英是由硅和氧组成的。石英在地球上有很多,而且很坚硬。因为这些特性,它用来做首饰已经有4000多年的历史了。在石英作为首饰之前,它用来作矛头。一些人认为石英有医治治疗的作用。

(4)欧泊文化:古希腊人相信,欧泊可让其所有者具有预见未来的能力。罗马人认为,欧泊是纯洁的象征。阿拉伯人认为,欧泊是天堂掉下来的。然而,到了19世纪,很多人认为,欧泊是跟坏运气连在一起的,是不可以佩戴的。现在仍有一些人这样认为。但是,现在很多人佩戴欧泊是因为欧泊很漂亮。

Unit 2　Optical Properties of Cut Gemstones
(琢型宝石的光学性质)

Part 5　Some Sentences (一些句子)

(1)宝石的美丽很大程度上取决于宝石的光学性质。最重要的光学性质是折射和颜色。其他光学性质包括出现斑斓的火彩、在不同方向宝石显示两种不同颜色的二色性以及透明度。

(2)钻石因其火彩和光泽而享有盛誉;红宝石和祖母绿因其亮度和美丽的颜色而珍贵;星

光蓝宝石和星光红宝石因为星光效应和颜色而广为人知。

（3）在一些宝石中，尤其是欧泊，在一个欧泊的不同区域能看到五彩缤纷的颜色，色调和色块还会随宝石的移动发生改变，这种现象叫做变彩。变彩不同于火彩，它是由于宝石内部的细小的不规则形状和裂隙对光的反射和干涉引起的。

（4）宝石的另一种光学性质是光泽，即宝石对光的反射能力。宝石的光泽可以用以下专业术语表示：金属光泽、金刚光泽（像钻石那样的光泽）、玻璃光泽（像玻璃那样的光泽）、树脂光泽、油脂光泽、丝绢光泽、珍珠光泽或土状光泽。宝石未切割时，光泽是其特别重要的鉴定标志。

Part 6　Short Paragraph（短文）

Optical Phenomena in Gemstones（宝石的特殊光学效应）

宝石的光学效应包含宝石随光线变化而具有的性质。这种性质不是由其化学性质或者晶体结构决定，而是由光与宝石内部特定的包裹体与结构特征相互作用决定的。宝石中的主要光学效应有晕彩效应、变彩效应、月光（光彩）效应、砂金效应、猫眼效应、星光效应和变色效应。

拉长石的晕彩呈彩虹色，是拉长石中的显微聚片（页片）双晶对光的薄膜反射引起的。

贵蛋白石中的彩虹色叫"变彩"。在欧泊中发生的不是散射，而是彩虹色。我们把所有的欧泊（一个宝石的大类）按其是否有变彩而分为宝石级欧泊和普通欧泊。

在月光石中，月光（光彩）效应是两种长石薄层交错使光发生散射所致。这些层的散射光均匀显示所有光谱色时呈现白色乳光，或者只呈现蓝色或蓝色和橙色，即为月光石的珍贵品种。像很多光学现象一样，层的大小和层间距影响所呈现的颜色。

砂金效应是光反射的结果。当另一种矿物以盘状或片状包裹体出现且其折射率都比较高的时候，它们就会像小镜面，使得宝石闪闪发光，这种闪光就叫砂金效应。最常见的反射物（包裹体）是铜、赤铁矿和云母。

猫眼效应也是由于光的反射，但这种效应不是由于片状包裹体的随即散射，而是针状或者管状包裹体像丝线一样平行排列对光的反射引起的。

星光效应是一种特殊的猫眼效应，在这些晶体中有几组定向平行排列的包裹体。与猫眼效应一样，不仅宝石内部的包裹体是定向的，而且切割呈弧面以显示其星光。

具变色效应的宝石在白炽灯和日光灯下其颜色明显不同。不管哪个宝石品种都能显示这种现象，但由于其在金绿宝石的变种亚历山大石中最强烈，因此，我们将这种现象称为"亚历山大石效应"。

Part 7　Text（课文）

Major Optical Properties of Gems（宝石的主要光学性质）

颜色

宝石的颜色由宝石选择性地吸收一些波长的光决定。我们知道，我们看到的白色光（或者无色）其实是由不同光组成的。

有色宝石的颜色从三方面描述：色调、亮度和饱和度。用这三个词可使宝石学家、珠宝商人与珠宝购买者描述和区分很多颜色并进行交流。我们逐一了解它们的含义。

色调：宝石的色调是它在光谱色中的基本位置，有红、橙、黄、绿、蓝、紫等。但它也包括各种过渡色,像微橙黄色或正蓝绿色。

亮度:宝石的亮度主要指颜色的明暗程度。它取决于色调,并在无色和黑色之间变化,很亮时看起来基本为无色,很暗时看起来基本为黑色。

饱和度:饱和度是衡量颜色纯净的程度,换句话说,就是相关灰色调或者棕色调出现的多少,宝石颜色的饱和度一般不进行量化,在很多情况下,只要色调和亮度相当好,颜色的饱和度就是宝石价值的主要决定因素。

光泽

宝石的光泽由宝石表面对光的反射能力决定。每种宝石和宝石变种都具有固定的光泽。而具体到单个宝石,其实际光泽都可能没有理论值那么强,这取决于宝石加工师的技术、宝石刻面状况、包裹体的多少或化学成分和物理性质的变化,如表面氧化或磨损会影响宝石表面的光泽。

通常以日常熟悉物体表面的光泽给各种宝石的光泽命名(前缀有 sub 表示"光泽弱于")。有的宝石的光泽十分具有代表性,就用这种宝石的名称来称呼这种光泽,如金刚光泽(金刚石就是希腊语中的钻石)和珍珠光泽。通览宝石学教材,你可以确认,绝大部分宝石都具有像玻璃一样的光泽或玻璃光泽。

宝石	光泽	宝石	光泽	宝石	光泽
黄铁矿	金属光泽	钻石	金刚光泽	锆石	亚金刚光泽
火玛瑙	玻璃光泽	萤石	亚玻璃光泽	软玉	油脂光泽
琥珀	树脂光泽	珍珠	珍珠光泽	虎睛石	丝绢光泽

透明度

宝石的透明度是宝石的非常直观而常见的性质之一。

宝石的透明度(或不透明)取决于宝石允许光透过的程度。宝石的透明度不但取决于宝石的化学性质和晶体性质,还取决于其厚度,与宝石的光泽一样,其透明度也与宝石的包裹体和表面特征有关。在下面的讨论和举例中,我们所说的是一种宝石常见的可能的最大透明度,而不是具体某个宝石的透明度。

当光照射宝石表面时,可能发生三种状况(这都和透明度相关)。所有的光会根据不同的比例发生反射、吸收和透射。这三个比例会决定宝石的透明度。

反射

观察反射时,光照射宝石表面和进入宝石内部都可能会反射回宝石或反射出宝石。

色散

色散(有时候也称"火彩"),是指将白色光分解为组成它的各单色光谱色,可能是红色、蓝色或绿色色斑,宝石转动时将闪烁着彩色光芒。

可以用数据精确地表示宝石色散值的大小,在实验室中用特殊的设备精心测试,计算出宝石红光下的折射率和紫光下的折射率之差即为色散。宝石的色散值一般为 $0.007 \sim 0.280$。

非实验室场合通常没有测试仪器,一般要用肉眼观察,简单的划分为:极低、低、中等、高、极高。宝石的色散强度取决于宝石品种(宝石的折射率),也与宝石的体色、宝石的切割比例有关。

具有低色散值的宝石有:萤石(0.007)、普通玻璃(冕牌玻璃或硅玻璃)(0.010)和石英(0.013)。不管颜色或切割如何,这些宝石的色散都很弱,基本上不可见。

具有中等色散值的宝石有电气石(0.017)、刚玉(0.018)和尖晶石(0.020)。这些宝石很少显示肉眼可见的色散,但有时候光照射在浅色、具有显著高冠部角的宝石上时,可能有色散。

具有高色散值的宝石有锆石(0.038)、钻石和蓝锥矿(都是 0.044)及立方氧化锆(一种合成宝石)(0.066)。这个范围内的宝石的色散通常是可见的。但体色较深的宝石和具有低冠部角的小粒宝石除外,其色散也是不可见的。

具有极高色散的宝石有闪锌矿(0.156)、钛酸锶(一种合成宝石)(0.190)以及合成金红石(0.280)。这类宝石很少有不显示明显的色散。

钻石是以高色散而著称的宝石,高色散也是钻石具有吸引力的品质之一。很多钻石的仿制品,无论是天然的还是人造的,是否能更好地仿钻很大程度上取决于其色散值与钻石的接近程度。

折射

SR 是单折射的缩写。这类单折射的宝石,不管光从哪个方向进入,在宝石中只有一个折射率(传播速率相同)。这类宝石包括非晶质的宝石材料,如欧泊、玻璃、琥珀等,以及立方晶系(等轴晶系)的晶体,最常见等轴晶系的宝石有钻石、石榴石和尖晶石。

DR 是双折射的缩写。一束光照射这类双折射的宝石时,将分解为两束相互垂直振动的光。这两束光以不同的速率沿着不同的(振动)方向穿过宝石晶体,有两个折射率,它们都是非等轴晶系的宝石晶体。

只有具有双折射的宝石才有双折射率,两束光的最高和最低折射率值之差即为双折射率,一般低至 0.003,高达 0.287。

当一个透明刻面宝石有高的双折射率,而且通过宝石台面的光不是光轴的方向时,光在宝石中轻微的不同步传播会造成宝石内部模糊或在粒度较大的宝石棱线边界看见有两个明显的影像,这就是常说的"刻面双影",这也是宝石切磨家头痛的问题,他们常常需要防止出现这个问题,必须找到晶轴方向作为宝石的台面。但这也常是肉眼或简单十倍放大镜下鉴定宝石时非常具有鉴定意义的特征。

多色性

多色性是从不同方向观察双折射宝石,宝石具有不同颜色或深浅不同的同种颜色的现象。

可将多色性分为弱、中或强。多色性的强弱取决于宝石品种和宝石颜色,也与具体某个宝石的色调有关。浅色宝石的多色性没有深色宝石的多色性强。除非效果特别明显(如堇青石和红柱石),一般切割好的宝石难以见到多色性,因为刻面使宝石产生的内反射色与棱线边缘颜色融合在一起,使多色性模糊了。

二色性宝石(如刚玉)呈现两种不同的颜色,而三色性宝石(如堇青石)呈现三种不同的颜色。

在单折射宝石中观察不到多色性,双折射宝石的光轴方向也观察不到多色性。

荧光

宝石吸收短波紫外线或者长波紫外线，或者同时吸收上述两种紫外线，宝石即刻发出可见光的现象叫做荧光。

我们把荧光不可见的宝石称为惰性宝石，宝石的荧光还可以呈现弱、中等、强等形式。宝石发出的荧光可能与宝石的体色相同，也可能不同。宝石在长波紫外线和短波紫外线下的荧光反应可能相同，也可能不同。

白色到浅黄色系列的钻石通常发出蓝色荧光，大约30%的钻石是这样的。

Unit 3　Physical Properties of Gems
（宝石的力学性质）

Part 5　Some Sentences（一些句子）

（1）贝壳状断口是既有特色又常见的一种断口，断口呈贝壳断裂面上的特征。这种断口在玻璃、石英、欧泊、橄榄石和琥珀等宝石中常见。其他可能的断口包括锯齿状断口、参差状断口和次贝壳状断口等。

（2）解理的知识对切割宝石的人来说非常重要，因为解理使得钻石精加工过程的第一步变得容易。说到彩色宝石，要避免解理面与宝石刻面平行，不然非常难以抛光。

（3）尽管比重测试可以应用于宝石原石和成品宝石，但是宝石必须是没有镶嵌的且由一种材料组成。不能测试镶嵌在首饰中的宝石或者组合宝石，如拼合石的比重；也不能测试多孔宝石的比重，因为它们吸收的液体会影响比重的测试精度；而且，在一些情况下，测试液会伤害宝石。

（4）具有一般或者较差韧性的宝石，其价值就不高。在保养、清洗和镶嵌这类宝石时要确保将其放在保护托中。

Part 6　Short Paragraph（短文）

Weighing Gems（宝石的质量单位）

在早期不同地区的珠宝市场上，常用当地一种或两种买家和卖家都熟悉的度量单位，如小麦种子或豆角种子表示宝石的质量或大小，类似这样物品的外形和质量都特别均一。现在，我们仍然可以从在宝石的国际公制单位中看到早期使用过的单位"克拉"，而"格令"这个单位有时候用于珍珠的交易，也用于作现代药剂单位。

克拉（carat）的英文发音跟蔬菜胡萝卜（carrot）的有点像，其缩写为"ct"，一克拉是0.2克，5克拉相当于一克。克拉是宝石商贸中的基本国际标准。对在英国或美国的人来说，英制单位更通用，如果要掌握克拉与克之间的关系，需要花费一定的时间并进行一些实践。盎司约相当于142克拉，是一个常见的英制质量单位。这样看来，真还没有适当的小英制单位可以容易地用于衡量宝石的质量。

另一个奇怪的美制单位是"开"，读音也跟蔬菜胡萝卜相似，缩写为"K"或"Kt"，指黄金的纯度。在大多数其他国家，黄金的纯度是指黄金的千分含量，如585或750，这样就不会与宝石的质量相混淆了。585就是合金中的黄金含量为585‰，也就是含金量为58.5%。

"开"制是将纯黄金含量分成24份，24K就是24/24，即纯金，18K就是18/24，14K就是

14/24（14K，18K，24K 换算成国际制分别是 585，750 和 999）。

Part 7　Text

Physical Properties of Gems（宝石的物理性质）

　　矿物的成分和晶体结构影响矿物的物理性质,使其具有各自的特征。宝石切磨者和镶嵌者都应该了解宝石重要的物理性质知识,消费者也可以借助这些知识来保存宝石。

相对密度

　　比重也被称为相对密度,不同的宝石具有不同的密度,是宝石鉴定中最重要的物理性质之一。相对密度（SG）是单位体积的宝石的质量与同等体积水的质量的比值。例如：蓝宝石（刚玉宝石）的相对密度为 4.0,意味着一立方英寸蓝宝石的质量刚好是一立方英寸水的质量的四倍。对天然宝石来说,相对密度值常常在一点多（琥珀为 1.08）到接近七之间变化（锡石为 6.95）。

　　有几种方法可以直接测量宝石的相对密度。测量相对密度时可以用重液。到目前为止,最精确的测量相对密度的方法是用一个特别改进过的天平测量宝石标本在空气中的质量（Wa）以及其在水中的质量（Ww）。根据阿基米德定律:"物体浸入水中的浮力等于排开同等体积水的重力",用一个简单的计算公式即可计算出非常精确的相对密度值。

硬度

　　矿物的硬度是指其抵抗刻划、研磨和切削的能力。矿物表面越能抵抗刻划,其硬度则越大,组成矿物的原子间的结合力就越大。常用摩氏硬度计来测试宝石材料的硬度大小。

硬度	1	2	3	4	5
矿物	滑石	石膏	方解石	萤石	磷灰石
硬度	6	7	8	9	10
矿物	长石	石英	托帕石	蓝宝石	金刚石

　　从上表可以看出,托帕石的硬度为 8,这意味着除了蓝宝石和金刚石之外,托帕石比其他宝石都硬。

解理和断口

　　解理和断口是指宝石遇到外力或压力发生破裂时所具有的特殊性质。有些矿物沿着平行原子结合力弱的方向有一种特别的裂开方式而形成一个平滑的表面,这种破裂称为解理。晶体矿物有解理和断口,而非晶质体或集合体矿物只有断口。

　　在矿物原石中,解理可以直接观察到,或者用锤子锤击矿物标本即可看到。钻石原石常劈开后再加工成形。在切割好的宝石中不大可能见到宝石的解理,除非宝石内部有可见瑕疵,或者意外地沿解理方向撞击使得宝石破裂。这样,尽管钻石是已知最硬的物质,但因其有较发育的解理可能使钻石遭受破坏。

　　断口是宝石破裂的一种方式。由于结合力在各个方向都差不多,它破裂的方向沿一定的方向而不是沿解理面。可以认为,断口类似于树木沿一个方向断裂,但这个方向不是其纹理的方向。

韧度或韧性
韧度或韧性是指石头经受外力或撞击而抵抗撕拉破碎的能力。容易被弄成小块或粉末的矿物被认为是脆性矿物。

压电性
有极轴的矿物或缺少对称中心的矿物受压力而产生电流即为压电性。晶轴在极轴的两端有不同的性质,当在晶轴两端施加压力时,电流流动可以产生正极和负极。石英和电气石具有压电效应。

导热性
一些宝石具有很好的热导性,如石英。当手握石英时,它可以把热量从人体中带走,因此触摸石英有凉感。不同材料因其晶系不同而有不同的热导性质。热导性不好的宝石,如琥珀,触摸其有温感,因为它没有把热量从人体带走。天然宝石的表面比玻璃和人造宝石的导热性要高得多。

在切割宝石的过程中,也要考虑导热性,因有些宝石在切割过程中需要一个冷却期。也可利用热导仪来区分钻石及其仿制品,因为钻石的导热性比仿钻要高得多。一些仪器用这个方法来鉴定其他宝石,但是十分昂贵,用时只有配合一些宝石学知识,小心谨慎才能判别宝石。

Unit 4　Gemological Instruments
（宝石鉴定仪器）

Part 5　Some Sentences（一些句子）

（1）当白光照射到宝石上时,一部分光从表面反射,而另一部分则进入宝石里面,会降低传播速率而发生"弯曲"或折射。这种光学现象就产生了一个非常重要的可测常量值,称为折射率。

（2）台式偏光镜可以快速而简便地检测宝石是单折射还是双折射宝石。

（3）棱镜式分光镜。用途：①未抛光的宝石;②鉴定经过处理的宝石;③折射率在折射仪刻度范围之外的刻面宝石;④鉴定一些合成宝石(如：区分天然蓝色蓝宝石与合成蓝宝石)。局限性：各波长不是线性均匀分布,红区收缩而蓝紫区扩宽。

（4）光栅式分光镜。用途：与棱镜式分光镜的用途相同。局限性：光谱不够明亮;难以控制进入仪器的光线;蓝区的光谱难以看到;优点：价格相对便宜。

Part 6　Short Paragraph（短文）

Different mensuration of gem's refractive index（测试宝石折射率的不同方法）

浸油是一种与众不同的测试宝石折射率的方法。将宝石材料浸到与其折射率相近的液体中,宝石将几乎不可见。将宝石浸到与其折射率不同的液体中,宝石将清晰可见或者明显凸现。这种方法用于检测未知矿物的折射率,需要将标本破碎,在显微镜下以观察其碎片在不同折射率的浸油中的状态。很显然,这种方法需要一套不同折射率的浸油以及一个高性能的显微镜,不适合于切磨或加工好的宝石。

浸油法有时不需要把宝石粉碎,只要简单地把宝石浸入"浸油瓶"或器皿中。将器皿悬置于一张白纸之上,器皿与白纸间留有空隙以便一张黑色卡可以从器皿底下穿过。透过液体和宝石从顶部观察,如果卡片边缘在一条直线上,则宝石的折射率和液体的折射率相同。如果宝石的折射率比较低,通过宝石看到的黑卡边缘就会比通过液体看到的黑卡边缘高;如果宝石的折射率高,则相反。

常见的浸油包括:水(1.33),普通酒精(1.36),丙酮(1.36),甘油(1.46),橄榄油(1.48),二甲苯(1.49),丁香油(1.53),溴化乙烯(1.54),三溴甲烷(1.60),二碘甲烷(1.74)。用重液测试比重时有时也可以看到宝石凸现。当宝石浸入到重液中,观察宝石的沉浮状态时,有时宝石却消失不见了,这种重液和宝石有着相似的折射率。

Part 7　Text（课文）

Gemological Instruments（宝石鉴定仪器）

用于放大观察的10×三合放大镜及显微镜

无论是用手持放大镜还是显微镜,其放大的目的都是使其表面和内部特征清晰可见。手持透镜或者小型放大镜都是单纯的透镜系统,可能有多种放大倍数。具三重消球差的放大镜是一种高质量的透镜,其外侧由两个无色含铅玻璃(或铅玻璃)透镜组成。(冕牌玻璃是一种由石英、苏打和石灰制成的普通瓶用玻璃;而无色含铅玻璃或铅玻璃是由石英、苏打和一种铅氧化物制成的,用来仿宝石。)尽管手持放大镜有很多种放大倍数,但大多数鉴定场合用10×(10倍)就足够了。放大倍数如果高于10×,就会产生宝石照明困难的情况,且视域和景深会变小。

如果要对宝石进行全面观察,双目立体显微镜是一种通用且十分有效的仪器。除了目镜之外,还有一个物镜,这样就产生两个透镜系统,可以提供一个精确的放大画面。这种显微镜的放大倍数就是目镜和物镜倍数的乘积(如10×物镜和3×目镜就是30倍的放大倍数)。宝石显微镜提供了一个"复反转"的图像(常规的复合显微镜会使图像反转),而且通常有缩放功能,可以连续改变放大倍数。顶灯是一个荧光灯源,而从底灯发出的光的光线很强,可穿过整个宝石。底光源配有一个可以调节光线强弱的可变电阻和一个控制从底部进入宝石的光线量的锁光圈。

用于测试宝石折光性的偏光镜

宝石材料可以分为两大类,各向同性或单折射宝石材料和各向异性或双折射宝石材料。各向同性或单折射宝石材料包括等轴晶系的晶体和非晶质材料,如玻璃、欧泊、石榴石和钻石。这些宝石有一个折射率,也就是说,光在这些宝石中发生平面偏振(在一个平面内发生偏振),在宝石的各个方向的传播速率一样。其他任何晶系的宝石矿物(四方晶系、斜方晶系、六方晶系,单斜晶系和三斜晶系)有双折射率值或者说是各向异性的。这些宝石材料有两个或者三个折射率,也就是说光沿晶轴的不同方向的传播速率不一样。当光进入一个双折射的宝石或各向异性的晶体时,光分解成两束在相互垂直平面上振动的偏振光。这两束光的传播速率不一样,可以用折射仪检测到两个折射率,而且能用偏光镜观察到。偏光镜是用来将宝石分为两大类的仪器,它由两个偏振片组成,一个起偏镜,一个检偏镜。

偏光镜可以快速检测宝石是单折射还是双折射。但是大多数宝石学家没有充分地利用它,像用锥光偏振仪来检测宝石的光性,是一个很好的检测宝石是单折射还是双折射的仪器。

但我们必须面对的是,要知道真正是单折射还是双折射最好使用折射仪。

用于测试宝石多色性的二色镜

　　光通过双折射宝石材料或者各向异性的宝石矿物时,光在不同振动方向的传播速率不一样,光在不同振动方向的吸收情况也不一样,因此颜色也不一样,这就形成了大家所知的多色性。如果宝石矿物中只有两个主振动方向的光,可以观察到两种颜色,就叫二色性。如果宝石矿物中有三个主振动方向(三个折射率),可以观察到三种颜色,就叫三色性(在任何方向上都只可以观察到两种颜色)。要观察多色性,宝石必须是:有色宝石(无色宝石显示的是组成白光的光谱色)、单晶(晶质集合体会使光散射,多色性模糊不清)、相对透明(大量的包裹体也会使光发生散射),以及观察方向非平行于晶轴方向。尽管用偏光镜可以观察宝石的多色性,但是冰洲石二色镜是观察宝石多色性的首选宝石仪器。这种二色镜是一个金属管,管的一端开口,另一端装有透镜。

　　光学冰洲石就放置在管里,形成两个四方形的观察孔。当把宝石置于明亮的灯源下,通过二色镜观察它,每个观察孔内图像都显示不同的颜色,这就表示是在正确的角度看到了不同振动方向的颜色,颜色波长也不同。当观察到了两种不同的颜色,要确定其多色性,可以将仪器旋转90度,这两种颜色在两个观察图像中会互换。可以通过改变观察的方向来观察宝石的三色性,观察到一种颜色后,在另一个方向可观察到另一种新的颜色。

用于测试宝石折射率的折射仪

　　用来测试折射率的仪器叫折射仪。折射率值可从数字刻度尺上读出,这个刻度线是阴影区和明亮区的分界线。这条线是观察可见光线以临界角照射进入宝石的一种测量方法,或者是光线射入宝石后可以完全反射回折射仪半球内,而不是反射或者折射出宝石。

　　折射仪的折射率上限可以读到1.81,由玻璃半球(放置宝石表面的工作台)和折射油(宝石和半球间的油,可以形成良好接触以避免外来的干扰)决定。要测试到最精确的数值需要用单色光源,或者说是钠光源。有一个测试值的宝石可能是单折射宝石,而有两个测试值的宝石可能是双折射宝石。当把可拆卸的偏光片放在折射仪的目镜一端时,转动偏光片(不是转动宝石),如果宝石的方位和光传播方位一致时,就可以看到双折射宝石的两个折射率。

　　反射仪可以用来测试任何折射率值的宝石材料,即使超过1.81也可以。

用于测试宝石发光性或荧光的长波或短波紫外荧光灯

　　宝石在短波或长波紫外光的照射下可以发射可见光,这种现象叫做发光,更确切的叫做荧光。在紫外光源撤离后,宝石仍发光的叫磷光。这种荧光或者磷光现象是由于其晶体结构内有杂质或者结构缺陷,吸收了紫外光,导致电子在能级之间振动,而发出的可见光。

　　当紫外线或不可见光直射宝石材料的时候,具有荧光性的宝石就会出现鲜艳的颜色。这个专业名词是从哪里来的呢?在19世纪中期,乔治·斯托克司教授用萤石在日光下做试验。他在一个暗屋中,让日光通过百叶窗上的一个小孔透进来,当日光照在一个无色萤石样品上时,他注意到矿物呈现一种明亮的蓝色。斯托克司教授把这种在萤石上观察到的现象新命名为荧光。他认为,这种光与欧泊中的乳白色光相似。因此,荧光这个词来源于萤石,但不是说所有的萤石都发荧光。荧光可以是深绿色、橙红色或者蓝白色,而且深浅不一。荧光可能是不可预测的,因为有些宝石在紫外光下没有反应。测试宝石的荧光性需要把宝石擦洗干净,而且要放在一个黑暗的地方观察。绝对不要直视紫外光,这样会对眼睛造成永久性的损伤。将宝石置于测试台面上以备测试（不要用金属钳子夹着或用手拿着）。看到宝石上的紫色为惰性的，这种紫色是光源反射的颜色。

用于测试宝石特殊发光性的滤色镜

滤色镜，也叫祖母绿滤色镜或者是查尔斯滤色镜，可以辅助鉴别一些天然宝石、合成宝石和仿宝石材料。颜色是宝石吸收和传播白光中的不同波长的光形成的。例如，绿色就是由不同波长的光混合形成的。这些不同的波长可有助于鉴别铬致色的祖母绿与其他因素致色的绿色宝石和玻璃仿制品。祖母绿滤色镜仅允许长波长的红光通过，吸收其他可见光，致使祖母绿在这种滤色镜下显红色。但这种测试对天然祖母绿不具有决定性，因为一些合成祖母绿和天然祖母绿在这种滤色镜下都没有反应。钴致色的合成尖晶石，普通的海蓝宝石以及托帕石仿制品在这种滤色镜下都显红色，但它们各自对应的天然宝石在滤色镜下都呈绿色调。染成蓝色的羟硅硼钙石和绿松石仿制品在滤色镜下显红色，但天然绿松石或合成绿松石却显绿色。

Unit 5   Synthetic Gems Introduction
（合成宝石介绍）

Part 5   Some Sentences（一些句子）

（1）可以在实验室合成出一些宝石如红宝石、蓝宝石、祖母绿和钻石等用以模仿天然宝石。

（2）最初在 1926 年人工生产出美丽的尖晶石，现在市场上已经有了很多颜色的合成尖晶石。其生产制造过程与合成刚玉的过程相同。由于尖晶石是等轴晶系的，没有二色性，据此可以把它与合成刚玉、天然刚玉或天然锆石及金绿宝石加以鉴别。

（3）劈开的梨形合成红宝石是顶级的无瑕合成刚玉宝石。这种材料中没有裂隙或瑕疵，通常是按劈开的梨形原石销售，而不是按克出售。一个劈开的梨形晶体一般是 28 克。这就很简单，即使最差的情况，梨形晶体也有 24 克，平均每克为 40 分币。

Part 6   Short Paragraph（短文）

Identification of Synthetic Gems（合成宝石的鉴定）

合成宝石在任何方面都与天然宝石十分相似。这包括具有相同的基本晶体结构、折射率、相对密度、化学成分和颜色以及其他特征。因为使用相同的宝石检测方法测试天然与合成宝石，有时候宝石学家也会迷惑宝石是天然的，还是合成的。如果碰到这种情况，最好的办法就是把宝石送到公认的宝石研究实验室，如美国宝石学院。他们可以准确地断定宝石是合成的，还是天然的。只有一些小的内部特征可以区别宝石是合成的还是天然的。

要区别天然与合成的无色蓝宝石是非常困难的。用显微镜可以看到在天然晶体中有不规则的气泡和液态包裹体，而合成宝石中沿晶面交叉处有裂隙。在显微镜下可见到有色合成宝石中有平行于梨形晶体顶部生长面的弧形生长纹。它们表示是不均匀的颜色分布。有时候，尤其是蓝色蓝宝石中，这种现象肉眼可见。

合成莫依桑石（碳硅石）

莫依桑石由金刚砂组成，其摩氏硬度为 9.25，是一种十分硬的物质。在陨石中发现过小的碳硅石颗粒，但是市场上当作首饰出售的碳硅石都是实验室合成的。

莫依桑石是用其发现者莫依桑的名字而命名的天然矿物。恒瑞·莫依桑（1852—1907）是1906年诺贝尔化学奖的得主。由于天然莫依桑石颗粒十分小，而且非常少见，因此不适合做宝石材料。莫依桑石（金刚砂）是在其天然品发现（1905年）之前合成成功（1893年）的。

　　因为合成莫依桑石的折射率为2.648和2.691，色散为0.104，摩氏硬度为9.25，相对密度为3.22，它比以往其他钻石仿制品的外观和硬度都更接近钻石。合成莫依桑石的导热性与钻石的导热性也很接近，用现在市场上的热导仪探测，其反应和钻石的反应一样。因为合成莫依桑石有较高的双折率，可以通过刻面重影来区分合成莫依桑石和钻石。其他的特征就是合成莫依桑石内有沿晶轴方向的平行排列的针状物和细小包裹体。

Part 7　Text（课文）

Synthetic Gems Introduction（合成宝石介绍）

　　合成宝石跟天然宝石十分相似，两者都具有相同的化学成分。合成红宝石和天然红宝石的化学成分都是 Al_2O_3，而且有相同的物理性质，如硬度、相对密度、解理和折射率等。但也有一些特定的偶然现象和特征可以用来辅助鉴别合成宝石和天然宝石。

　　尽管开始的过程十分艰辛，但在19世纪后期，随着焰熔法合成红宝石的发展，合成宝石开始走向成功。随着工业上对各类高品质晶体的应用（如：光学配件、激光晶体等其他用途）需求和对晶体生长机制的逐渐知晓，已经生长了各种晶体，一些还是天然宝石的合成对应产物。合成宝石材料与天然宝石矿物有相似的化学成分和晶体结构。相反，宝石的仿制品与天然宝石有相似的外观，但化学成分、物理性质和晶体结构却不一样。上述两种合成宝石材料都有各自的宝石学特征，可以借此将其与天然对应品区分开来。

　　现在合成宝石的方法分为两类：从不同组分的流体中结晶（例如熔融法或水热法）以及从大概与晶体化学成分相似的熔体中结晶。合成钻石是在高温高压下在金属熔体中生长的，直到20世纪70年代早期首次合成出宝石级金刚石晶体才受到珠宝首饰商贸界的关注。然而，由于其生长条件、昂贵的费用以及有限的可用生长设备所限，实际上宝石级的合成钻石一直很有限。那些出现在珠宝交易中的合成钻石主要是棕黄色的、重约1克拉或更小的晶体（1克拉＝0.2克），磨成刻面成品重约0.5克拉或更小。而与合成钻石相反的是，无色或近无色的仿制品材料大量充斥市场。

　　这些年来，大量的天然宝石材料和合成材料用来仿无色钻石。立方氧化锆（立锆或CZ）是使用最广的仿制品，因其成本低，且其外观与抛磨好的钻石相似。可以通过一种简单的宝石测试仪器测量它的热导性，将其与钻石区分开。在过去的两年中，一种新材料，合成莫依桑石（金刚砂）作为首饰进入市场。它给鉴定带来了一定的麻烦，就是因为不能采用上述提及的测试其热导性的方法来区分它与钻石。然而，合成莫依桑石因其具有各向异性的光学特征而具有特征的光学性质（放大检查可以见到双影），这一特征易于被受过训练的宝石学家鉴别出来。

　　彩色宝石中，最重要的合成宝石是合成刚玉（蓝宝石和红宝石）、祖母绿、尖晶石和紫晶。如前所述，它们可以用溶液法或熔融法合成。在珠宝市场上，焰熔法和晶体提拉法的生产成本较低，因而市场上大量出现这两种方法生产的宝石，而助熔剂法或水热法生产的产品却比较少。从溶液中生长紫晶属于相对便宜的那一类，因为电子工业生产合成水晶的大型生产设备是现成的。与之不同的是，水热法或者助熔剂法生长红宝石、蓝宝石和祖母绿却是非常"奢侈的"合成方法，需更高的价格。

合成红宝石和蓝宝石的方法:1902年法国化学家奥古斯特维尔纳叶将氢氧火焰倒置于一开口的陶瓷马弗炉中,最先开始从细粒氧化铝中合成出梨形的晶体(与刚玉具有相同物理与化学性质的氧化铝粉末)。这个方法稍加改进就可以用来生产尖晶石、金红石和钛酸锶。

超纯氧化铝粉末置于底部有一个精细筛子的容器中,如果容器受到由机械制动锤子的敲击,氧化铝粉末就会掉入封闭的炉体腔中。氧气进入炉体时,携带了精细的氧化铝颗粒进入到炽热的氢氧火焰中间,在这里它们熔融成液滴滴落在梨形晶体的表面。火焰参数、粉末进入的速率以及梨形晶体下降的速率都调整好,以生产一个直径一致的梨形晶体。梨形晶体表面的温度刚好高于熔点,对无色蓝宝石来说是2030℃(约3690°F)。当梨形晶体达到所需尺寸(一般为150～200克拉)时,就关闭炉子,冷却梨形晶体。

在冷却的过程中,会产生应变,因为表面的冷却速度比内部的要快,这种现象会产生裂隙,造成合成过程中大量的损耗。这种应力可以通过纵向劈开梨形晶体来释放。一些残余应力对宝石来说没什么不好,大部分工业用途的材料通过劈开选用梨形晶体的一半。梨形晶体可以在1950℃以下退火释放应力。

Unit 6　Gemstone Treatments and Enhancements
（宝石的处理和优化）

Part 5　Some Sentences（一些句子）

(1)"优化"这个专业名词是指宝石除了切割与抛光以外其他的改变外观(颜色/净度/现象)、耐久性和实用性的任何改善过程。

(2)许多人把宝石改善与合成、人造、实验室培养或其他方法制造在地球上不存在的材料的过程等专业术语弄混了。

(3)事实上,大部分用做首饰的宝石材料都是经过处理改善其外观的。处理的宝石材料是一种很好的选择,因为它们比未处理的宝石材料有更好的质量。

Part 6　Short Paragraph（短文）

Symbols for Specific Forms of Enhancement（优化方法的特定代号）

B＝漂白:用热、光或化学方法,或其他试剂来弱化或消除宝石的颜色。通常伴随这个方法之后的是染色或注入。例如:漂白养殖珍珠;漂白/注入翡翠(B货翡翠)。

C＝覆膜:通常指采用上漆、上釉、涂墨、衬底、电镀等改变表面外观,增加颜色或其他特殊效果的优化方法。如:钻石覆膜。

D＝染色(着色):把染料导入到宝石材料中以给宝石一种新颜色,或增强宝石现有的颜色,或改善宝石颜色的均匀度。如:染绿色翡翠。

F＝充填:在宝石表面的空洞或裂隙中注入无色玻璃、塑料或一些类似的物质。这个过程可以改善宝石的耐久性、外观或增加其质量。如红宝石。

Fh＝熔融愈合处理:在热处理过程中,熔融液可能用来愈合早期形成的开放型裂隙或裂缝,这个过程使破裂处的周围发生溶解,熔化的宝石材料再沉淀,使宝石的裂隙愈合。如:红宝石(尤其是缅甸孟宿红宝石)。

H＝热处理:这是用加热使颜色、净度或特殊现象发生改变。如:红宝石、蓝宝石、坦桑石、海蓝宝石和翠榴石。

　　I＝注入：注入是往多孔宝石中加入无色物质（通常是塑料），以增加宝石的耐久性和改善其外观。如：充填绿松石。

　　L＝激光：用激光和化学药剂去除并改变包裹体。如：钻石。

　　O＝浸油/树脂充填：在宝石表面裂隙中充填除玻璃或塑料之外的物质，如无色油、蜡、树脂或其他无色物质，以达到改善宝石材料外观的目的。如：祖母绿。

　　R＝辐照处理：用中子、γ射线、紫外线或电子轰击来改变宝石材料的颜色。辐照之后常有一个热处理过程。如：蓝色托帕石。

　　U＝晶格扩散处理（体扩散或表面扩散）：深度扩散是用有色的化学试剂经由高温热处理来产生颜色或星光效应。如：晶格扩散处理蓝宝石。

　　W－上蜡/上油：往多孔宝石材料中注入一种无色蜡、石蜡或油以达到改善外观的目的。如：翡翠。

Part 7　Text（课文）

Gemstone Treatments and Enhancements（宝石的处理和优化）

　　宝石材料的优化处理方法的出现已经有数百年的历史了。最早发表宝石处理文章的是普林尼。2000多年后的今天，许多方法仍在使用。有些宝石优化方法是改进宝石的自然状态，持久且不能被检测出来，这给宝石市场提供了很多美丽的宝石材料。其他处理方法可以给宝石本身或宝石的净度带来明显的变化，如用辐照和热处理的方法使无色托帕石永久性地变为蓝色托帕石就是一个很好的例子。少部分处理方法不够稳定，掌握宝石知识的买家应该避免购买这种宝石。接下来将介绍一些常见的处理方法。这仅仅是宝石改善方法的冰山一角。

　　在过去，宝石商为了他们的成品宝石价值尽量高，通常让宝石切磨者对宝石进行处理。现在，有一些专攻宝石处理的中心用设备对宝石原石和成品宝石进行处理，如泰国曼谷。热处理刚玉宝石（红宝石和蓝宝石）是一个非常好的例子。刚玉的热处理（包括简单的热处理和添加铍的热熔融处理）通常在切割之前进行，而且在切割之前宝石商是不知道的。

宝石处理与价格

　　有一些宝石在自然界不曾存在，是处理之后才出现的。大量的各种黄色、金色和橙色的黄水晶是紫晶经热处理后的产品。天然的黄水晶非常稀少。如果没有经过处理，这种宝石会比现在贵很多。

　　市场上各种蓝色或紫色的坦桑石也是热处理的产品，极大地满足了人们的需求。

　　粉红色的托帕石是另一个不经热处理就很难得到的宝石品种。这些处理方法不但被人们所接受，而且能充分保证市场供应。

　　近年来，对未经热处理的蓝宝石和红宝石的需求，使得未经热处理原料的价格上涨50%～100%。这是否意味着未处理的宝石就更漂亮呢？其实并不是这样！在大部分情况下，热处理过的宝石更漂亮，但是因为未经处理的宝石稀少，所以价格更高。

热处理

　　热处理是最常用到的处理方法。它可以减弱、加深或完全改变宝石的颜色。它可以改善宝石的净度和亮度。热处理的宝石只有训练有素的鉴定师在实验室条件下才能检测出来，而热处理在常规条件下是不可改变的。实验室常通过检测红宝石和蓝宝石中含有微小的金红石针或液态包裹体中的小气泡以确定其未经热处理。如果这些宝石的颜色很好,又因为其非常稀有,其价格一定很高。

下面这些宝石通常都是经过热处理的:坦桑石、黄晶、粉色托帕石、海蓝宝石、帕拉伊巴碧玺(备注:1989 年发现自巴西 Paraiba 产出的绿蓝色—蓝色调的电气石)、磷灰石、红宝石、蓝宝石、锆石(包括蓝色和无色的)。

浸油

祖母绿中浸油是普遍现象,但不是每一个祖母绿都是浸油的(上极品祖母绿可值天价)。把开采出来的祖母绿原石放入油桶中,当工匠切割祖母绿时,油就充当了润滑剂。无色油渗入到祖母绿表面的裂隙中。如果裂隙中含有油,则就很难观察到裂隙。为了完成这个过程,要把油压入到抛光好的祖母绿裂隙中。这个方法的过程是可接受的。你可发现祖母绿没有浸油的唯一方法是祖母绿表面没有裂隙,则没有油能进入其内部。很显然,如果祖母绿的颜色相同,表面又没有裂隙,包裹体更少,则其价格更高。如果把最初有裂隙的祖母绿放入超声波或蒸汽中清洗,其中的油就会渗出裂隙。这可能使表面的包裹体显得更白且更明显。在这种情况下,祖母绿可以重新浸油。

最近,有些文章报道:其他颜色的宝石,如红宝石、变石、金绿宝石的其他变种以及翠榴石都被注入了油或树脂,以使其表面包裹体不显眼。有时在祖母绿和红宝石中也注入有色油。这么做的目的就是使裂隙更隐蔽。购买时要避免买到这种宝石,因为你看不到这种宝石的真实颜色,也不知道裂隙有多大。所幸的是,这种宝石不常见。在美国,如果购买有明确来源的祖母绿,基本上碰不到这种注有色油的宝石。祖母绿和其他表面有裂隙的宝石也在其裂隙中注入了合成树脂。在这个过程中加入硬化剂使得宝石更持久。这种加入树脂和硬化剂的方法是不可接受的。

辐照

辐照就是用亚原子颗粒或放射线轰击宝石材料。有时候辐照之后再加热处理可使宝石产生一种更好的颜色或新的颜色。蓝色托帕石就是一个最常见的例子。尽管也有天然产出的蓝色托帕石,但是非常少,而且带灰色调。美国的放射性宝石都是由放射性管理机构监管,必须没有有害的放射性残留才能进入市场。

但在美国之外的国家就没有这个保证了。现在,可以辐照出天然蓝色托帕石中未曾出现的蓝色调托帕石,其价格相当合理,因为批发市场上有大量的竞争产品。如果你能找到一个未经处理的蓝色托帕石,其价格可与未处理的帝王托帕石匹敌。通过辐照,可以将碧玺处理成深粉色—红色,与天然的红色碧玺难以区分。颜色不好的钻石可以经辐照和热处理成深绿色、黄色、蓝色、棕色和粉色。这些宝石相当常见。同等颜色、净度级别和大小的天然彩钻比辐照彩钻的价值高很多。辐照可以使养殖珍珠呈灰色或蓝色,而通常这些颜色是染成的。辐照珍珠和染色珍珠的价格差不多,比上好颜色珍珠的价格低很多。各种石英和锂辉石经过辐照后再经退火热处理可以产生所需的艳丽颜色。

染色

如果没有染色,就不会有黑色缟玛瑙,天然的玉髓也没有这种颜色!玉髓,更常见的是玛瑙,常常被染成蓝色、绿色或橙色,雕刻成碗、雕像或珠子。这下好了,因为有许多可爱的作品用这个材料,尤其是雕刻成动物之类的,没有人会在乎这是不是天然的。在日本阿古屋母贝中养殖的珍珠可达 10mm 以上,长成具有某种颜色且有伴色的精品。如果是灰黑色、浅蓝色、紫色、近黑色或深青铜色,就可以认为是染色的。为了满足市场对玫瑰色珍珠的需求,一些养殖珍珠被染成粉色,这可以通过仔细观察染色珍珠的孔和色斑来辨别。另一方面,南海养殖珍珠通常比日本养殖珍珠要大,可以形成具有很多奇异天然色彩的珍珠,因为它们是在不同的贝中养殖的。

漂白和覆膜

漂白是对有机宝石材料的处理方法,如象牙、珊瑚、珍珠和养殖珍珠。这种方法可以使颜色变浅,而且持久、难以鉴别。不存在与天然颜色产品之间的价格差异。

覆膜处理已有两百多年的历史了,它是用油漆或者薄层来改变宝石的外观。现在,覆膜还进一步用来改变宝石的颜色。虹彩托帕石就是 Azotic 公司在托帕石表面使用覆膜技术的例子。这个公司现在使用覆膜技术将托帕石处理成各种颜色,包括粉红色和各种艳丽的"帝王"色调。最近有报道指出,对坦桑石的亭部进行覆膜处理,用来增加其颜色的饱和度。鉴定时,偶尔见到在颜色不好的钻石表面有覆膜改善其颜色外观,用来欺骗消费者。

可能在欧泊的底部覆上一层黑色膜,以增加变彩并使欧泊表现为黑色体色。可以采用覆一层黑漆或叫"欧泊二层石"的方式,将一个欧泊薄层胶合在黑色玉髓基底上,就是"欧泊二层石"。

扩散

扩散处理最早用来处理蓝宝石。在高温下,将化学试剂如铍渗入宝石内。早期的扩散处理只是在宝石表面形成颜色,叫做"表面扩散"。早期时,表面扩散在浸油中用放大镜很容易鉴别。在最近十年,扩散处理方法取得了非常大的进步,研究发现如果刚玉在超高温中恒温较长时间,扩散就可以渗透到整个宝石中。

扩散处理可以增强宝石的颜色,改变宝石的颜色,或者产生星光效应。

充填

充填用于宝石的表面裂隙和孔洞中。玻璃、塑料或其他物质都可用来当填充材料。有时候也会充填红宝石。放大近距离观察可以看到充填部分与宝石表面的光泽不同,或者在暗域照明下,在裂隙中看到闪光效应。亚洲宝石研究院(AIGS)曾对充填红宝石做过深入的研究。

充填钻石

有时候用玻璃充填有包裹体的钻石使其看起来更纯净。奥德钻石和吉田钻石都经过了充填处理。加热、超声波清洗或重新翻新都可能破坏填充物。充填并不是修整包裹体,而是使其不显眼。

在灯下近距离观察钻石,旋转钻石,可以见到蓝色的闪光。如果吉田法和奥德法处理的钻石受损,可以免费重新填充。购买此种钻石之前要仔细核实保证书。

激光处理

有时候用激光技术处理钻石。激光打孔就是在钻石上打一个非常小的洞直至影响钻石美丽的包裹体。如果这个包裹体不能被激光熔蚀,就需要对这个包裹体进行蒸发或漂白。在正确的角度,通过放大检查可以看到激光孔。不管激光打孔钻石看起来外观的纯净度如何,它必须定级为有微瑕疵或有瑕疵,而且价格也要相应降低。

未优化的宝石

有一些宝石还未见对其优化,包括:石榴石(除翠榴石之外)、橄榄石、堇青石、尖晶石、各种金绿宝石、碧玺(除帕拉伊巴碧玺之外)、孔雀石、赤铁矿和长石(除中长石和拉长石之外)。始终要牢记的是:宝石材料的处理技术日新月异,有许多改善方法即使都可能,也很难鉴别开来。

Unit 7　Gemstone Inclusion
(宝石包裹体)

Part 5　Some Sentences(一些句子)

(1)由于天然宝石和合成宝石中包裹体的形成机理不一样,因此它们的特征也各有不同,也可以利用包裹体来区别不同产地的宝石。

(2)宝石改善方法,如表面扩散、体扩散、铍扩散处理等,只能通过浸油观察宝石才能鉴别。非专业人士将水和婴儿用油当成浸油放在器皿中,而专业人士用一种和宝石折射率相近的液体作为浸油来测试宝石。

(3)包裹体提供了鉴别宝石母岩的重要信息,有时候可以揭示宝石的来源,有时候可以显露天然宝石经过处理的迹象,有时候可以通过包裹体来区别仿天然宝石的合成宝石和仿制宝石。

Part 6　Short Paragraph(短文)

Example of Inclusions in Diamond and Emerald
(示例:钻石和祖母绿中的包裹体)

钻石包裹体是钻石内部的特征,因为有包裹体就意味着钻石不完美,故也叫瑕疵。

包裹体就像指纹一样,可以展示宝石的所有特征信息。知晓你的钻石内部和外部特征,可以使你对自己的私有财产更了解。如果万一宝石丢失或者被盗,这将有助于你描述和鉴定自己的宝石。

一些包裹体会影响钻石的净度,会影响通过钻石的光线,使钻石看起来没有那么明亮,其他类型的包裹体会使钻石更容易破碎。

只有非常少量的钻石是完美的,而这些完美钻石非常昂贵,所以我们购买的钻石或多或少内、外部都有点瑕疵。大部分珠宝商告诉我们,如果钻石的包裹体没有影响钻石的强度或者严重影响外观,就不要担心这些包裹体。

几乎所有的天然祖母绿都含有包裹体,它们在鉴别天然祖母绿和合成祖母绿及其他绿色宝石中非常重要。有些典型包裹体具有产地特征。结合折射率及相对密度,包裹体可以提供祖母绿出产国的鉴别依据。

看上面的图:大部分祖母绿都含有大量的裂隙和缺口。这张照片显示了哥伦比亚祖母绿中的典型三相包裹体,这个负晶形的三相包裹体包含了一个气泡、一个石盐晶体和一份盐水溶液。如果把这种包裹体切开,露到表面,包裹体内就会充满空气,形成反射。这种包裹体常被注油或注树脂以减少反射。

拼合宝石中,冠部都用天然宝石材料。

Part 6　Short Paragraph(短文)

Gemstone Cut Shapes(宝石的切割形状)

圆形

这种形状已是所有钻石切工的标准形状,现今,圆形切工的钻石占钻石销售量的75%以上。其58个刻面是通过公式精确计算切割而来,可以获得最好的火彩和最大亮度。

公主形

是一种由众多闪闪发光的方形或长方形小刻面构成的切工,一种相对较新颖的切工,而且,发现多被用于单粒镶嵌作订婚戒指。适合长手指的人佩戴,两侧多用三角形宝石点缀。

祖母绿形

一种去角的长方形切工。据说称作阶梯状切工是因为它的台面、四周的平滑切面像楼梯的台阶。

椭圆形

这是一个均匀的完美对称设计,深受手小或手指短的女性青睐。长圆型能产生手指修长的美好感觉。

梨形

这是一种混合切工,是椭圆形和马眼形的最好结合,很像一滴闪光的眼泪。这种设计也属于钻石切工大类,最能弥补手指较短或中等长度的缺陷,作吊坠或耳坠特别漂亮。

垫子形

由一个底部小切面与几个大的刻面、一个敞开的底面和一些圆形角组成。切工深度吸引目光透视宝石内部,刻面可增加光线内反射的程度,使宝石具有最好的亮度和钻石切工火彩。

心形

是一种最基本的爱的象征符号,是一个顶部开个口的梨型钻工。加工者的技能决定切工的漂亮与否,务必寻找一块形状匀称和轮廓鲜明的心形切工宝石。

三角形

宝石具有三角形刻面,通常很闪耀,由此而得名。三角形切工的基础是43个刻面,但现在已经变成了50个刻面或更多刻面。由于为等边三角形,故有充足的光线和颜色进入视野。三角形切工几乎与圆钻形切工一样明亮,因此,对于那些喜欢有圆钻的明亮度又不喜欢圆钻的购买者是一个很好的选择。

弧面形

是对宝石表面进行精抛光、凸面切工、手感光滑、无刻面的切工类型。

Part 7　Text(课文)

Gemstone Cut(宝石加工)

尽管可以找到一些好的晶体,但大多数宝石都经历了水蚀作用损坏或呈鹅卵石型。因为宝石的光学性质可以产生最佳效果,所以适当的刻面型切工能使宝石具有更高的价值,也更加漂亮。宝石的加工旨在修正晶体使之匀称。

宝石的切割质量可以对它的外观产生重大影响，而对克拉价格影响很小。珠宝商和精明的珠宝消费者都会为珠宝的美丽付出代价，特别注重宝石加工。

你怎么能够讲出宝石加工的好坏呢？最简单训练眼睛的方法是先看差的加工再看好的加工。图示的三个红宝石，颜色和净度都很好，中间那颗红宝石看起来更好，因为所有射入宝石的光线都返射回人眼。查看颜色是否均匀地分布于宝石表面。这颗红宝石也有生气和闪光，如同光线在刻面上跳跃。相反，另外两块宝石有暗色区域，光线没有沿适当的角度全部返回观察者的眼睛。

加工款式也可以增加宝石的美丽程度。看看这两块宝石：它们的大小、形状、品质和颜色都相同，而外形却显示出明显的不同，因其中一个是标准的祖母绿切工，另一个是重子切工，背面有更多的小刻面。

除了这些标准的宝石切割形状外，珠宝设计师们针对个性化款式正在发明一些新的宝石切工方式，如一些具有不寻常几何造型的刻面宝石、雕刻宝石、刻面与雕刻相结合的宝石。

一个好的切工能够展示宝石的颜色、减少包裹体并呈现良好的整体对称性及切工比例。由于宝石的颜色会变，因此，当谈到最好的火彩或颜色时没有硬性的几何学标准。宝石，尤其是稀有宝石，有时其切割只关注保重而不考虑颜色因素。例如，刚玉的变种宝石蓝宝石和红宝石，加工时会尽可能保重优先于切割漂亮程度，此时可能会出现色带或生长纹。

对一个颜色较深的宝石，其最好的切工可能是比一般宝石切得浅一些，允许更多的光线入射到宝石中；而对颜色较淡的宝石，切得厚一些可能对颜色更有利。

查看镶嵌的宝石，确保所有刻面都是对称的。一个不对称切工的冠部表明宝石的切工质量低。在所有情况下，切工好的宝石都是对称的，反射光均匀地穿过表面，且抛光光滑，没有任何刻痕或擦痕。像钻石一样，颜色级别高的宝石通常都有台面、冠部、腰棱、亭部和底尖（面）。变彩欧泊是个特例，绝大部分为圆弧型切工。

大多数不透明的宝石，包括欧泊、绿松石、缟玛瑙、月光石等，一般都加工成弧面型，而透明宝石一般加工成刻面型。你也会看到品级较低的蓝宝石、红宝石和石榴石加工成弧面型。如果宝石材料的颜色非常好但不够透明或洁净以加工成刻面宝石，那么也可将其加工和抛光成非常诱人的弧面型。因摩氏硬度小于7的宝石容易被粉尘和研磨砂里的石英划伤，则较软的宝石材料也通常加工成弧面型，弧面型宝石上显示的微细划痕比刻面型宝石上的要少得多。

Unit 10　Diamond
（钻石）

Part 5　Some Sentences（一些句子）

（1）简单说来，钻石越大，越珍贵（以克拉质量衡量）；钻石越纯净，越珍贵（以净度衡量）；钻石的色泽越浅，越漂亮（以色泽衡量）；钻石切割越精确，越能光泽熠熠（以切工衡量）。

（2）克拉是钻石的重量计量单位，也是4C中最易鉴定的，但两颗一样大小的钻石其品质却有不同，因为钻石品质还取决于色泽、净度与切工。

（3）"钻石恒久远，一颗永流传"是全世界男女老少都熟知的，这句话对每个人又意味着不同的含意。

(4) 净度是钻石纯净度的标志。除了在极少数情况下,微量的矿物、气体或其他元素在钻石结晶时被包裹于其中,称为内含物,是大自然遗下的印记。它们看起来像是小晶粒、云彩或羽毛,也正是它们使颗颗钻石均不同,每一颗都焕发出独特的神采。

Part 6　Short Paragraph（短文）

Diamond Grading and the 4C's（钻石的 4C 分级）

大家知道,4C 是构成钻石价格的基础,净度、颜色、切工(比例)和克拉质量共同构成钻石质量要素,决定钻石的价值。切记,每个"C"都对钻石质量起重要作用,例如,一粒无色钻石为顶级颜色,但如果它不够洁净、很小或切工不够好,那么它的价值就会比较低。最好的钻石拥有 4C 中每个 C 的稀有品质,也非常有价值。

净度

钻石的净度由包裹体即内含物的数量、种类、位置、大小、颜色和瑕疵即表面性质所决定。所有钻石都具有识别特征,但绝大多数肉眼不可见。为了观察钻石,专家通常使用 10× 放大镜检测微晶、羽状裂隙或云雾状等外观。根据分级师观察包裹体和瑕疵的难易程度最终将净度分为 5 个级别。

LC(镜下无瑕):内部和表面无瑕的钻石最稀少且最漂亮。

VVS(极微瑕):含有极小的包裹体,即使是一个娴熟的钻石分级师在 10× 放大镜下检查也很难查看到。VVS_1:极难观察,只能通过亭部才能观察到,非常细小和不明显,重新精抛光可去除;VVS_2:很难观察。

VS(微瑕):含有较小的包裹体。VS_1:一个训练有素的分级师在 10× 放大镜下难以观察;VS_2:一个训练有素的分级师在 10× 放大镜下较容易观察。

SI(瑕疵):含有明显的包裹体。SI_1:10× 放大镜下容易观察;SI_2:10× 放大镜下很容易观察,有些肉眼可观察。

P(重瑕疵):一个训练有素的分级师在 10× 放大镜下能明显观察到包裹体。常常从台面方向观察,肉眼可见包裹体,严重影响钻石的耐久性,或由于瑕疵太多而影响其透明度和火彩。

颜色

在理想情况下,钻石根本没有颜色,像一滴泉水。随着颜色由浅变深,钻石颜色级别从无色(D)到深色(Z),超过"Z"级、颜色浓艳的钻石称为"彩钻"。已知色级的钻石常被用于色级比对分级。颜色分级就是未知色级钻石与已知色级的钻石,即"比色石",进行比对,可以在人工光源或北半球的日光下进行比对。一种被称作"色度计"的仪器可被用作颜色分级,但其不能代替训练有素的分级师的眼睛。

克拉质量

克拉是所有宝石的质量单位。一克拉分为 100"分",因此,一颗 75 分的钻石是 3/4 克拉重,或为 0.75 克拉。一克等于 5 克拉。"克拉"这个词来自热带豆槐的种子。直到这个世纪,这些种子的质量一直作为衡量贵重宝石的称重单位。

切工

宝石学家所说的"切工"是指成品宝石的比例,如亭深、台宽和小刻面的均匀度等所有制约火彩、耐久性和其他的钻石外观特征。

切工在确保钻石重量的同时,要尽可能提高钻石的亮度和火彩。切工实际上指两个方面:首先是它的形状(圆形、椭圆形等);第二是如何很好地切磨钻石。

Part 7　Text (课文)

Gemological Characteristics of Diamond(钻石的宝石学特征)

在珠宝贸易中,钻石是所有宝石中最重要的品种。据估算,钻石销售占全球珠宝贸易额的近90%。钻石总是被加工成小刻面以展示其独一无二的光泽和火彩。它至高的硬度确保其切工的精度,是所有宝石中唯一的。

一般将钻石分为工业级和宝石级两种,后者通常用"4C"分级标准:克拉质量、净度、颜色和切工。除了标准圆钻型外,钻石也常被加工成梨形、椭圆形、心形、马眼形、三角形、长方形或祖母绿琢型。通过精心的设计,钻石可被镶嵌到所有精美的首饰中,如项链、耳饰、戒指等。钻石的宝石学特征如下:

(1) 化学成分:C(碳)。

(2) 晶系:等轴晶系。

(3) 晶体习性:最重要的是八面体。钻石也呈立方体、菱形十二面体、似立方体等。钻石晶体经常畸变,晶面可能弯曲。可有八面体晶体(双晶常见)。

(4) 表面特征:可见八面体晶面上的三角形蚀象。解理:完全的八面体解理,可用于钻石抛光、开料或修整有瑕疵的钻石材料,可见于成品钻石和钻坯的内部及表面。

(5) 摩氏硬度:10。钻石是已知自然界物质中最硬的。钻石的硬度依结晶方向不同而改变。但在任何方向,钻石依然比其他宝石硬得多。

(6) 比重:3.52。

(7) 颜色:无色、浅黄色、褐色或绿色。彩色(具有明显色调的那些)包括黄色和褐色,少见有绿色、粉色和蓝色,非常罕见的是红色和紫色。除了彩色钻石外,钻石的价值随着色调的加深而降低。

(8) 光泽:金刚光泽。

(9) 折射率:2.42,单折射。许多钻石具有异常消光。

(10) 色散:高,0.044,钻石的色散值比其他任何天然无色宝石都高。

(11) 发光性:钻石的荧光变化范围较大,从无到强,荧光颜色多变。长波紫外光下的荧光强度比短波紫外光下的强,为蓝白色到紫色、绿色或黄色荧光。也有些钻石呈现荧光惰性。钻石的荧光变化可用于鉴别群镶的无色钻石。如果一件首饰上所有的宝石都呈现相似的荧光特征,则这些宝石就不太可能是钻石。

紫外灯下呈蓝色荧光的那些钻石可能具有黄色磷光,这是鉴别钻石的诊断性依据。

(12) 产状:主要产于金伯利岩管或冲积矿床中。

(13) 产地:印度的冲积矿床是自古典时代到18世纪间已知的唯一钻石矿资源。巴西的钻石矿藏大概发现于1725年。

在19世纪后半叶发现了南非冲积矿床和金伯利岩管钻石矿床,西伯利亚钻石矿床发现于20世纪40年代。近来,澳大利亚成为了重要的钻石产地,产于与金伯利岩相似的钾镁煌斑岩中。

宝石级钻石的重要生产国有安哥拉、澳大利亚、博茨瓦纳、巴西、中国、纳米比亚、俄罗斯、塞拉利昂、南非和坦桑尼亚等。

(14) 相似宝石：许多天然的、合成的及人造品被用于仿钻石。其中外观最相近的有：

① 合成立方氧化锆(CZ)；

② 人造钇铝榴石或 YAG；

③ 无色锆石；

④ 某些玻璃。

到目前为止，合成立方氧化锆(CZ)是最好的和应用最广泛的仿钻品。其他用作仿钻的有天然和合成的白色蓝宝石、合成白色尖晶石。

Unit 11 Ruby and Sapphire
（红宝石和蓝宝石）

Part 5 Some Sentences（一些句子）

(1) 定向排列的金红石包裹体产生六射星光效应，成为受欢迎的星光红宝石。

(2) 蓝宝石以其蓝色为人熟知，但它几乎可呈任何一种颜色。

(3) 针状金红石是红宝石或蓝宝石中一种常见的包裹体，当其定向排列及加工正确时可产生猫眼效应或星光效应。

(4) 泰国是全球最重要的红宝石交易中心。或许某个时候，一个交易周期内，全球80%的红宝石都在泰国进行买卖。最大的红宝石切磨工厂在泰国的尖竹汶府，曼谷是世界红宝石的集散地。

Part 6 Short Paragraph（短文）

The Color of Ruby Gemstones（红宝石的颜色）

哪种颜色会自然而然的使你将爱、活泼、激情和力量联系起来呢？这是显而易见的，不是吗？红色。红色是爱的颜色，她散发出温暖和强烈的生命力。红色也是宝石之王红宝石的颜色。红宝石是表达热烈感情的一种完美方式。配有珍贵红宝石的首饰能使人感受到彼此的热情和炙热的爱。在红宝石的颜色中，最明亮和最好的颜色被称作鸽血红。真正的鸽血红极其稀有，它产于缅甸抹谷矿区，具有超越物质世界、超越想象的颜色。一位抹谷商人最好地表述，他说"求见鸽血红红宝石就像求见上帝的脸"。

数千年来，红宝石被认为是地球上最有价值的宝石之一，它拥有一粒珍贵宝石应有的一切：华丽的颜色、极好的硬度和显著的光泽。除此之外，它相当稀少，尤其是优质红宝石。

颜色是红宝石最重要的性质，其次为透明度。因此，包裹体不会降低红宝石的质量，除非包裹体降低了宝石的透明度或正好位于红宝石台面的中心。相反，红宝石内的包裹体可被称作"指纹"，一种特征标识，同时也是真正天然红宝石的证据。切工是根本：在某些情况下，只有完美的切工才能彰显珍贵红宝石的美丽，才是显示"宝石之王"的一种恰当方式。然而，一块完美的红宝石就像完美的爱情一样稀缺。如果你偶尔遇到它，那是你的运气。但当发现属于你的红宝石时，千万别犹豫，佩戴上它！

Part 7　Text（课文）

Gemological Characteristics of Ruby and Sapphire
（红宝石和蓝宝石的宝石学性质）

红、蓝宝石是宝石级刚玉矿物，它们都属于五大名贵宝石中的品种。红宝石是七月的生辰石，是结婚15周年和40周年的纪念石。蓝宝石是九月的生辰石，是结婚5周年和45周年的纪念石。红宝石和蓝宝石的宝石学特征如下：

（1）化学成分：Al_2O_3，铝的氧化物。

（2）晶系：三方晶系。

（3）结晶习性及表面特征：六方柱状或六方板状晶体，另有平行双面单形。在各单形交替面角处可见菱形面，还会出现六方双锥，显示不同程度的长而细和短而粗的双锥形。有时刚玉呈现桶状外观。锥面和柱面体上具有垂直C轴的晶面横纹。在底轴面上常见三角形生长标志，这是非常好的识别特征。

刚玉通常产于水蚀鹅卵石中，表面可见双晶面。

（4）解理：无解理，具有贝壳状或参差状断口，具有平行底面和菱面体的裂理。

（5）摩氏硬度：9。

（6）相对密度：4。

（7）光泽：玻璃光泽到亚金刚光泽。

（8）折射率：1.76～1.78。

（9）双折射率：0.008。

（10）光学性质：一轴晶。

（11）颜色和品种：纯净的刚玉无色，常见红色、蓝色、粉色、橙色、黄色、绿色、紫色和黑色。宝石级红色刚玉就是红宝石，其他颜色的宝石级刚玉都称为蓝宝石。

（12）多色性：红宝石——强多色性，深红色和橙红色；蓝宝石——其他颜色比黄色的多色性强，其他颜色蓝宝石的多色性会显示不同的体色色调。

（13）光谱：红宝石——由铬致色，诊断性吸收线为红区有一条吸收双线和两条弱吸收线，黄绿区全吸收，蓝区有3条细的吸收线，紫区全吸收。手持分光镜一般只能看到蓝区有两条吸收线。红区的双线和偶尔见到的其他线是作为明亮的发射线而不是吸收线被观察到。蓝宝石——蓝色蓝宝石的吸收光谱通常是在蓝区有3条吸收带，在许多绿色和金色蓝宝石中也可见到相似的吸收光谱。

（14）包裹体：刚玉内的固态晶体包裹体有磷灰石、方解石、绿泥石、刚玉、萤石、石墨、橄榄石、黄铁矿、石英、尖晶石、电气石和锆石等。其他包裹体还有液态、气态和固态，充填于空洞内形成了微细图案，这可能会与合成刚玉中的熔融包裹体混淆；生长带为由直线和角构成的六边形，反映了刚玉的晶系和晶形；熔蚀的固体包裹体有针状金红石和盘状赤铁矿—钛铁矿。金红石包裹体被熔蚀成"丝"并在底轴面三维方向以60°或120°的夹角定向排列。

（15）产地：许多地方都产出商业级红宝石，最主要有：缅甸、越南、巴基斯坦、阿富汗、泰国、柬埔寨、斯里兰卡和坦桑尼亚。蓝宝石的主要产地有斯里兰卡、澳大利亚、克什米尔、缅甸、泰国、柬埔寨和美国。

（16）发光性：红宝石、粉色和紫色蓝宝石都含铬，因此，在紫外灯下呈红色荧光。天然蓝宝石常含微量铁,铁会抑制发光。

(17) 星光红宝石和星光蓝宝石:具有好的体色和清晰星光的天然红、蓝宝石很稀有。通常当存在大量、沿晶体横向结晶轴定向排列的针状金红石包裹体时,可见六射星光,少见十二射星光。宝石必须切割成弧面型,包裹体平行于成品宝石的底面。天然星光宝石所含的针状金红石通常比合成星光刚玉中的粗一些。

(18) 合成刚玉:可以用多种方法合成各种颜色的刚玉,不同的合成方法可以产生不同的特征包裹体。

(19) 人工处理:绝大多数红、蓝宝石都经过热处理以提高其颜色级别。用表面扩散法对蓝宝石和低档红宝石进行处理,用以改善或增强颜色以提高卖相。可用红色的油充填红宝石的裂隙,以提高其颜色和净度级别。可用玻璃充填红宝石中的裂隙、空洞和龟裂纹。

Unit 12　Emerald
（祖母绿）

Part 5　Some Sentences（一些句子）

(1) 泰国和印度是祖母绿的主要加工中心。美国和日本是最大的祖母绿消费国,几乎占全球购买总量的75%。

(2) 大粒的祖母绿很稀有,因此,祖母绿的每克拉价格也随着宝石大小的增加而大幅度上涨。

(3) 有色宝石中透明度和净度紧密相联,尤其是祖母绿。祖母绿通常都含有可见包裹体。因此,销售带有包裹体和瑕疵的祖母绿得到了市场的认可。

(4) 祖母绿是最难加工的宝石之一,加工人员应非常小心,在他开始每个过程前都要计划切工。天然祖母绿的内部具有许多裂隙(或裂缝)使之易脆。

Part 6　Short Paragraph（短文）

General Information of Natural Emerald Gemstone（天然祖母绿简介）

祖母绿这个词使人联想到乐园美景中的青葱绿色。千百年来,人们热爱、欣赏天然祖母绿宝石的翠绿色——自然美的象征,因此,虽然她具有脆性,也很难镶嵌到首饰上,但她仍然是最流行的宝石之一。祖母绿也是五月生辰石。

祖母绿的名字起源于波斯语(Esmeralde),历史上已知最早的天然祖母绿宝石产于红海附近的埃及,埃及祖母绿矿就在斯科特山和扎巴拉赫山的山坡上。后来,因哥伦比亚开采出优质的祖母绿,埃及矿就失去了它的重要地位(几乎被遗忘了几个世纪)。大且珍贵的祖母绿矿石也被用于制作祖母绿牌匾。

顶级祖母绿产自哥伦比亚。迄今为止,哥伦比亚祖母绿在全球珠宝市场上以难以置信的价格出售。祖母绿也产于印度、赞比亚、巴基斯坦、阿富汗、俄罗斯、南非、埃及、津巴布韦、奥地利、巴西、澳大利亚、坦桑尼亚和马达加斯加岛。

事实上,据说哥伦比亚祖母绿被发现时,人们宁愿选择承受痛苦的折磨甚至是死亡也不愿意开采祖母绿矿石资源,这就是产于哥伦比亚的美丽、高贵的祖母绿宝石。

在远古时代(约公元前4000年巴比伦时期,已知最古老的珠宝市场),最好的祖母绿宝石晶体被奉献给女神维纳斯，代表不朽的声名与忠贞的信仰，这就是为什么祖母绿能成为订婚，甚至结婚戒指的原因之一，它与白色的结婚礼服形成很好的对比。

Part 7　Text（课文）

Gemological Characteristics of Emerald（祖母绿的宝石学性质）

祖母绿，绿柱石矿物的绿色亚种，是最知名和最受欢迎的绿色宝石。它美丽的绿色、耐久性和稀有性使之成为最有价值的宝石之一，顶级的祖母绿甚至比钻石更有价值。

祖母绿的化学式为铍铝硅酸盐（$Be_3Al_2(Si_2O_6)_3$）。绿柱石也包括其他鲜为人知的宝石品种，如海蓝宝石和金绿柱石。祖母绿的相对密度为 2.67～2.78；折射率值为 1.56～1.57 和 1.59～1.60；摩氏硬度为 7.5。虽然硬度相对较高，但其脆性也很大。纯净的绿柱石是白色的，祖母绿的绿色由杂质离子铬（Cr）（偶尔也为钒）所致。

市场上，祖母绿的颜色决定其价格，但切工、净度和宝石大小也是价格的决定性因素。深绿色祖母绿已被作为所有其他绿色宝石的颜色标准。但在该领域各个行家对祖母绿的颜色问题也有争论。

虽然颜色较淡的品种确切地应该称为绿色绿柱石，但传统上，任何一个铬致色的优质绿色绿柱石都被称为祖母绿。不过，即使珠宝实验室已经有比色石可将祖母绿从绿色绿柱石中区分开，但对什么颜色为浅色（绿色绿柱石）还是有分歧。

有人可能想知道，什么颜色的祖母绿最受青睐。答案就是蓝绿色—绿色且微量元素含量变化在铬、钒和铁之间，它们当中任何一个元素的存在或缺失都将导致祖母绿产生色差。

最常用的祖母绿优化处理方法是浸油，伴随浸油也有许多填充物。天然祖母绿晶体被浸于有色或无色油或树脂中一段时间。经过多次加热后以至于油从宝石裂隙中渗入。这种方法可帮助填充祖母绿的裂隙，使其看起来包裹体较少，颜色也得到改善。

按照贸易惯例，尽管浸油处理的祖母绿是被认可的，但用有色油、树脂或任何其他充填物处理过的祖母绿出售前都应该声明，同时价格也会大打折扣。

早期的科学家已经多次在实验室里合成祖母绿宝石，但首次具有商业价值的合成祖母绿产品大约在 1940 年由卡罗尔·查塔姆完成的。到 1961 年，经奥地利的约翰·李奇来尼进行了产品改进，由林德公司推出了"林德合成祖母绿"。1000 多克拉的大祖母绿晶体由查塔姆和后来的吉尔森生产出来。一般在市场上可获得 5 克拉以上切割的成品祖母绿宝石。

合成祖母绿或实验室生产的祖母绿的市场价格通常比天然的低得多。依据质量以及与天然祖母绿的相近程度，合成祖母绿的价格为每克拉 1～150 美元之间。合成祖母绿总有很好的透明度和最好的绿色。用肉眼很难鉴别合成祖母绿与天然祖母绿，如有疑问，最好将祖母绿送到权威珠宝实验室进行检测。

有些低档宝石可能与天然绿色祖母绿混淆，如：绿色绿柱石、翠榴石、绿色锆石、绿色蓝宝石、碧玺、玉髓、东陵石、合成祖母绿和萤石。

Unit 13　Quartz Gemstone
（石英类宝石）

Part 5　Some Sentences（一些句子）

（1）血滴石是一种不透明的、深绿色的玉髓，其上面分布有铁的氧化物引起的红色斑点。

（2）某些宝石，特别是石榴石、托帕石、橄榄石和碧玺，属于化学成分相关联但各样品又不尽相同的矿物族。因此，在这些宝石中，各样品的折射率和相对密度等性质都有所不同。

（3）碧玉属于玉髓（微晶石英）类，不透明，颗粒细小。碧玉可呈红色、黄色、绿色、灰蓝色、褐色及其组合色。碧玉含有一定量的有机质。

（4）虎睛石石英含有褐色铁质而产生金黄色，加工成素面宝石可显示猫眼效应（表面有小的光束），看起来像老虎的眼睛。虽然虎睛石在西澳大利亚、缅甸、印度和美国加州也有发现，但最重要的矿床还是在南非。

Part 6　Short Paragraph（短文）

The Purple Quartz Crystal——Amethyst（紫色石英晶体——紫晶）

紫晶，石英家族里最受欢迎的宝石，也是昔日珍贵、稀有、具有皇室优雅的宝石。由于全球都能发现大量高质量的紫晶晶体，现今已经变成具有魅力、可买得起的宝石了。紫晶是二月的生辰石。这种精美的宝石因希腊词"Amethustos"而得名，意为"不醉"，源于相信佩戴这种宝石不会因饮酒过量而遭受惩罚。紫晶晶簇呈漂亮的紫色小晶体，为全球收藏家所喜爱。

据历史记载，紫晶曾经非常受欢迎，以至于在市场上紫晶与红宝石、祖母绿或蓝宝石同等重要和昂贵。实际上，在公元前2500年的法国，新石器时代的人类就用紫晶作装饰品。早在公元前3100年，在埃及和希腊，紫晶被制成珠串、首饰和印章制品。

在整个欧洲，一直到1799年，紫晶才赢得高贵的评价。1799年，在俄罗斯的乌拉尔山脉发现了一个储量丰富、高质量的紫晶矿床并开始开采，这些紫晶矿导致紫晶宝石的价格下降。

在中世纪的欧洲，紫晶宝石被认为在战场上能够保护士兵。古代希腊人相信紫晶具有魔力，这些美丽的宝石用来入药。

深受欢迎的紫晶首饰包括紫晶耳饰、耳环和耳钉。由于它们的颜色和厚重的形状，紫色的紫水晶被制作成订婚戒指和结婚戒指。

Part 7　Text（课文）

Quartz Gemstone（石英类宝石）

石英是地球上最常见的矿物之一，被珠宝界广泛认知。石英既漂亮又耐久，价格也不贵。它可被加工和雕刻成各种形状和大小的成品。

石英主要有两个变种：显晶质石英和隐晶质石英，其化学成分相同，都为二氧化硅（SiO_2）。显晶质石英包括无色水晶（无色）、紫晶（紫罗兰色）、黄晶（金黄色）、玫瑰水晶（粉红色或桃红色）、烟晶（棕褐色）、东陵石、石英猫眼和其他品种。这种石英几乎都是透明到半透明的，隐晶质石英，显微小晶体，如玛瑙、水草玛瑙、火玛瑙、血滴石、绿玉髓、绿石英、硅化木、碧玉、红玉髓（红玛瑙）、苔藓玛瑙、缟玛瑙、缠丝玛瑙、肉红玉髓等。隐晶质石英通常呈不透明或半透明。

石英的摩氏硬度为7，折射率为1.544~1.553，双折射率为0.009；三方晶系，无解理；透明，玻璃光泽；吸收光谱和荧光性因品种不同而异。

在珠宝文化中，无色水晶和烟晶曾被用于水晶球，被算命先生、女巫或吉卜赛阿婆用来给人预知未来。紫晶是二月生辰石，而黄水晶是十一月生辰石。

购买石英时，深色者最有价值。一块好的石英应该透明、无包裹体。由于晶体中颜色分布不均匀，石英常被加工成明亮式圆钻型以最大限度地展示其颜色。当颜色分布较好时可见用其他琢型。

石英内无包裹体时,透明石英的价值最高,而另外一些则主要因其中的包裹体才有价值。最受欢迎的是含金红石针的发晶。金丝发晶是透明的无色水晶,内部含有定向排列的金色丝状金红石。每一块发晶都不同,且有些发晶出奇地漂亮。包裹体有时被称作维纳斯毛发。一个鲜为人知的品种是电气石石英,电气石替代了金色金红石,包裹体是黑色或暗绿色电气石晶体。

无色水晶都未经处理。有时通过染色(如玛瑙)、辐照(低放射性轰击)或加热等处理得到有色石英宝石,用以改善颜色。诚实的珠宝商人总会告知顾客他们所用的优化处理方法。

石英矿床产地:

无色水晶:阿尔卑斯山、巴西、马达加斯加岛、美国;

烟晶:巴西、马达加斯加、俄罗斯、苏格兰、瑞士、乌克兰;

紫晶:巴西、玻利维亚、加拿大、印度、马达加斯加、墨西哥、缅甸、纳米比亚、俄罗斯、斯里兰卡、美国(亚利桑那州)、乌拉圭和赞比亚;

黄晶:阿根廷、巴西、马达加斯加、纳米比亚、俄罗斯、苏格兰、西班牙、美国;

玫瑰石英:巴西、印度、马达加斯加、莫桑比克、纳米比亚、斯里兰卡、美国;

东陵石:澳大利亚、巴西、印度、俄罗斯、坦桑尼亚;

石英猫眼:巴西、印度、斯里兰卡;

鹰眼石:巴西、印度、斯里兰卡;

虎睛石:澳大利亚、印度、缅甸、纳米比亚、南非、斯里兰卡、美国。

Unit 14　Pearl

(珍珠)

Part 5　Some Sentences (一些句子)

(1) 天然珍珠是完全由自然过程产生的,而养殖珍珠则开始于人为的有意干扰继而养殖而成。

(2) 最重要的产珠软体动物是海生贝类和淡水蚌类。

(3) 珍珠属于有机宝石材料,是最古老的珍贵宝石之一。历史上,珍珠曾非常珍贵,仅次于钻石。

(4) 珍珠相对较软,遭受刮擦、受热和化学物品侵蚀等容易受损。但正确保管,珍珠会陪伴你一生或更长时间。

(5) 与天然珍珠相比,养殖珍珠相对便宜。高质量的直径 3mm 的养殖珍珠大约每克 10~25 美元。价格主要取决于形状、光泽及表面完好度。

Part 6　Short Paragraph (短文)

Natural Pearls (天然珍珠)

成分:天然珍珠由 80% 以上的碳酸钙、10%~14% 的介壳质和 2%~4% 的水组成。碳酸钙为斜方晶系的文石,方解石可能存在于某些淡水珍珠内。

比重:2.60~2.78。

折射率:1.530~1.685。

摩氏硬度:3.5~4。

结构:天然珍珠具有由介壳质胶结文石的碳酸钙层围绕中心显微小核而形成的同心圈层状结构,每一个圈层代表一个生长期。

碳酸盐圈层内,各晶体的 c 轴呈放射状排列,此特性具有重要的宝石鉴定意义。文石晶体呈片状和叠瓦状,就像屋顶上的瓦片。

光泽:独特的珍珠光泽受复杂因素影响,这与抛光宝石的表面光泽区别很大。珍珠表面对光的反射和微细的虹彩形成珍珠光泽。虹彩由下列原因产生:

a. 文石叠瓦式晶片边缘的衍射作用;

b. 光穿透文石薄晶片时发生干涉,然后又反射回珍珠表面。

反射、干涉和衍射的综合效应就形成了珍珠光泽,这不是一个按科学方法定义的术语,而是描述一种微妙效应,以区别于体色。

评价珍珠的质量取决于观察者的经验。

颜色:按照体色大致将珍珠分为白色、粉色、黄色、灰色、青铜色、绿色、紫红色和黑色。另外,光泽本身可具有柔和的但以粉色、黄色、绿色或蓝色为主色调的颜色。关于灰色和黑色珍珠的呈色原因已有多种理论依据,包括贝类分泌珍珠质里的一些矿物和蛋白质,以及珍珠表面附近富介壳质层的存在。

形状:理想情况下为圆球体,但许多珍珠达不到这种形状。梨形(或水滴形)的珍珠也有很高的价值,扁球状被称作纽扣珍珠。极不规则的珍珠被称为(异形)巴洛克珍珠。

天然珍珠的有限供给已经激起人类试验在人类的帮助下刺激贝类生产珍珠的兴致。中国从 13 世纪就开始尝试养殖珍珠了。

Part 7　Text（课文）

Pearl（珍珠）

珍珠,宝石皇后,一直为人们所喜爱。珍珠分成天然珍珠和养殖珍珠。天然珍珠的形成开始于没有任何人类协助而异物进入贝类体内,受异物刺激的贝类为了摆脱刺激而分泌出被称为珍珠质的柔滑物质层层包裹异物,多年后,珍珠层越来越厚,最终产生出美丽而光亮的珍珠。

为了养殖珍珠,有必要模仿大自然。小心地撬开贝类壳,将异物放入其体内,因受刺激该贝类分泌出珍珠质。接下来珍珠自然生长,在合适的时候就形成了光彩的珍珠。海水珍珠与淡水珍珠间存在着差异。淡水养殖刺激物为蚌类的外套膜片,而海水养殖刺激物为珠核的贝壳小球粒。养殖珍珠早已达到了人类的意愿。珠宝商致力于养殖珍珠源于极似天然珍珠的养殖珍珠的质量和价格有很大的竞争优势。随着过渡养殖和水域污染,除了拍卖行和古董店,天然珍珠已相当罕见。

选择珍珠时要考虑五个特征:光泽、质地、形状、大小和颜色。

光泽:光泽是许多珍珠质层产生的漂亮珠光。珍珠质层越厚,其珍珠光泽就越好。光泽是珍珠分级最重要的因素。

质地:因珍珠是天然形成的,几乎不可能找到没有瑕疵的珍珠。瑕疵越小,则珍珠的质量越高。

形状:珍珠很少能达到精圆形或近圆形,常常为不规则形。珍珠越圆,其价值越高。

大小:珍珠的大小用直径来表示,范围为 1~20 毫米。珍珠越大,其价值越高。

颜色：可能有蓝色、黑色、金色、粉色和橙色的珍珠，但最常见和最受欢迎的仍然是白色或乳白色珍珠。

产地：过去最重要的珍珠养殖业场位于波斯湾和马纳尔湾（位于斯里兰卡和印度之间）。这些水域的养殖工作仍在继续，但由于水域污染和经济环境的改变产量大大减少。在澳大利亚北部外海海域，珍珠母贝的养殖得以恢复。其他海域包括太平洋诸岛、红海、委内瑞拉和墨西哥海湾。

大多数养殖珍珠业集中于中国和日本。另外澳大利亚、缅甸、菲律宾、塔希提岛、泰国和其他国家也养殖珍珠。

Unit 15 Jadeite Jade
（翡翠）

Part 5 Some Sentences（一些句子）

（1）术语"玉"表示两种不同的岩石，它们分别由不同的硅酸盐矿物组成——硬玉和软玉。

（2）目前，虽然大多数宝石按克拉质量出售和估价，但翡翠通常是按件出售。

（3）像其他精美宝石一样，也有人对翡翠进行改善甚至是合成的尝试。目前的珠宝市场上仍然没有用于商业的合成翡翠。

（4）传统的翡翠鉴定方法主要依靠折射仪、宝石显微镜、查尔斯滤色镜和分光镜，所有这些方法在鉴别漂白翡翠（俗称B货）时都非常困难。

（5）在中国，翡翠原指一种鸟名。翡是一种红色羽毛的鸟，翠是一种绿色羽毛的鸟。后来翡和翠渐渐融和在一起特指一种带褐色的蓝绿色小鸟。

Part 6 Short Paragraph（短文）

Chinese Jade Culture（中国玉文化）

中国的玉文化博大精深，玉象征着美、高尚、完美、坚韧、力量和不朽。中国人爱玉不仅因为她的美丽，更因为她的文化、象征意义和灵通，就像孔子所说的玉有十一种美德。玉文化为中国文化所特有，正像中国人常说的"黄金有价，玉无价"。

玉的历史与中国文明一样源远流长。考古学家已经发掘出新石器时代早期的玉器。新石器时代早期文化以浙江省河姆渡遗址为代表；新石器时代中晚期文化以辽河流域的红山文化、黄河流域的龙山文化和太湖流域的良渚文化为代表。

因玉代表着漂亮、优雅和纯洁，她被用于中国许多成语或俗语中，代指美丽的人或美好的物，如玉洁冰清（纯洁与高贵）、亭亭玉立（举止文雅、苗条和优美）、玉女（漂亮女孩）。中国特色的玉常被用于人名。

玉被制成祭祀器皿、工具、装饰品、器具和许多其他物品。古代乐器就用玉制成，像玉笛、玉箫（直立的玉笛）和玉钟。古时候玉对中国人来讲也充满神秘，因此，当时很流行用玉制器皿作祭祀用品和陪葬。统治者中山靖王刘胜（公元前113年）为了保存其尸体，死后埋葬时穿戴用金丝线将2498块玉片缝制在一起的金缕玉衣。

Part 7　Text（课文）

Jadeite Jade（翡翠）

宝石级翡翠为多晶质集合体的岩石，是著名的两种玉石材料之一，另一种是软玉。翡翠集高强的韧性和精美漂亮于一体，是理想的雕刻、珠串加工材料。翡翠（翡翠的主要组成矿物）的宝石学特征如下：

(1) 化学成分：$NaAl(SiO_3)_2$，钠铝硅酸盐。

(2) 晶系：单斜晶系。

(3) 习性与产状：用于雕刻和作宝石的翡翠是多晶质集合体。它产于变质岩中，也以块状和鹅卵石产于河流冲积矿床中，具有粒状、交织的韧性结构，颗粒内的交织纤维颗粒镜下可见。

(4) 硬度：大约为7，不同方向略有差异。

(5) 比重：3.30～3.36。

(6) 颜色：白色、淡紫色、紫罗兰色、红色、橙色、黄色、褐色、淡绿色到翠绿色、典型的绿白斑杂色、深绿色到黑色。翡翠卵石可呈现几种颜色的杂色；风化的外皮可产生棕色"皮"。最为珍贵的品种为透明到近透明的祖母绿色，被称为"帝王玉"。

(7) 吸收光谱：以蓝/紫区有强吸收线、蓝区伴有弱吸收带为特征。翠绿色翡翠因铬致色而具有典型的铬吸收谱——红区有双吸收线，有时蓝区有一条吸收线。其他绿色和棕褐色翡翠由铁致色，淡紫色翡翠由锰致色。

(8) 光泽：油脂光泽到玻璃光泽。

(9) 透明度：透明（非常稀少）、半透明到不透明。

(10) 折射率：1.65～1.67，通常在折射仪上仅能在接近1.66的地方看到一条模糊的阴影边界。

(11) 双折射率：双折射晶体的无规则取向导致翡翠呈半透明，表现为在正交偏光镜下各个方向都有光（不消光）。

(12) 产地：商业级翡翠的主要产地为缅甸，另外，日本、美国加利福尼亚、危地马拉和俄罗斯也有产出。

(13) 加工：弧面、珠形、雕件和首饰用小挂件。在早期的翡翠成品以及没有用金刚砂抛光的翡翠上，用放大镜清晰可见不均匀抛光特征，即"橘皮效应"。不均匀抛光是由于岩石中晶体颗粒不同方向的硬度微差异引起的。用金刚石研磨剂抛光能产生光滑且具有强光泽（玻璃质）的表面。

(14) 处理：通常要经过漂白、着色、充填和其他处理方法等。翡翠仿冒品包括半透明的祖母绿、玉髓、鲍文玉和其他蛇纹石玉、玻璃和塑料。

Unit 16　Jewelry Commerce
（珠宝商贸）

Part 5　Some Sentences（一些句子）

(1) 生辰石是代表接受者出生月份礼物的宝石，托帕石是传统的11月生辰石，据说佩戴托帕石可使佩戴者受益。

（2）评估宝石的外观，检查项链和手链的爪扣，核实戒指和耳坠上镶嵌的宝石，如果金属廉价而易坏，宝石很可能不贵重。

（3）现代优质宝石一般镶嵌在铂金或至少14K金中。一些珍贵宝石镶嵌在标准白银中，但这种情况比较少见，一般是将半宝石镶嵌在标准银中。

（4）大多数宝石按其重量标价，但一些雕件和弧面型宝石例外，它们按件而不是按克拉出售，因为制作成本超过了材料的价格。但对绝大多数宝石，价格还是以克拉计算。

（5）很多人不知道蓝宝石可以是无色的，钻石可以是黄色的，石榴石可以是紫色的，托帕石可以是粉色的。颜色虽不是宝石分类中最突出的因素，但其仍然可以影响宝石的价格。紫色或绿色的石榴石、红色或粉色的托帕石、红色的钻石都是最稀有很珍贵的颜色类别。任何宝石，具有鲜活而纯正颜色者都是最珍贵的。

Part 6　Short Paragraph（短文）

Gemstone Certificate（宝石证书）

美国宝石学院钻石报告能说明钻石的真实性，常用于美国和世界其他国家。报告详尽地描述影响钻石质量、美观和价值的每一个关键评估指标。

一般来说，宝石证书可以保证宝石的真实性。有证书的宝石在市场上很容易售出。现在，我们来看看以下3个证书，学习专业鉴定参数。

Part 7　Text(课文)

The Price of Gemstone(宝石的价格)

不同的宝石每克拉的价格可能差距非常大，理论上1克拉宝石的价格变动范围为一美元到数万美元。影响每克拉宝石的价格有许多因素，这里简要总结10个影响宝石价格的因素。

宝石品种

一些宝石品种，如蓝宝石、红宝石、祖母绿、石榴石、坦桑石、尖晶石和变石，由于它们的优质宝石特性和稀缺性，在市场上售价颇高。其他种类，比如各种石英，在世界很多地方大量存在，其价格就很低。宝石的种类大致决定了其价格范围，每块宝石的特性在很大程度上也影响每克拉的价格。

颜色

对有色宝石而言，颜色是影响价格最重要的因素。理想的颜色因宝石的种类而异，但一般浓、鲜、纯是最重要的。太深或太浅颜色的宝石在价格上会比适中的同类宝石便宜。所以矢车菊蓝的蓝宝石会比墨水蓝或浅蓝的蓝宝石的价值高得多。

净度

一块非常干净、没有包裹体的宝石的价格相对高一些。一般来说，宝石越干净，就越光芒夺目。诚然，宝石净度级别越高，其价值就越高，但有些不影响宝石火彩和闪耀的包裹体不会太影响宝石的价值。同时要知道，有些宝石，如祖母绿，总是会含有包裹体的。

切割和抛光

宝石应该按正确比例切割，以最大限度地将射入宝石的光返回到人眼。但宝石工匠或珠宝商在切割具体宝石时，常常不得不做些妥协。如果宝石的颜色太浅，切得深一些可以获得更丰满的颜色。相反，将深色宝石切得浅一些可淡化成品宝石的颜色。但无论哪种情况，宝石的刻面应该整洁，表面应该被很好地抛光，没有划痕。

大小

对某些宝石品种，比如石英类宝石，不管其重量多少，每克拉的单价相对稳定。但对稀有宝石品种，价格并不随着重量的增加而线性增长。实际上如某些宝石，比如钻石，随着宝石克拉数的增加，价格呈现幂指数增长（暴涨）。按这个模式，一个 1 克拉的宝石其价格可能是 1000 美元，而两克拉的就需要 4000 美元。尽管这个模式并不那么确切，但优质的红宝石、蓝宝石越大，就越具有每克拉价格更高的趋势。

不仅越大的宝石其价值越高，而且切割成珠宝商贸中所知标准尺寸的宝石也会更贵。因为为了达到那个标准，更多的部分会被切掉。

琢型

某些琢型宝石会比其他琢形宝石的价格高一些，部分是因为需求，部分是因为切割成特定琢形时材料方面的问题。市场上一般圆形成品宝石的价格会高一些。圆形宝石比椭圆形的不常见一些，因为椭圆形切工的宝石可以保留更多的材料来增加重量。将宝石切成圆形会损耗更多的原石，对贵宝石比如红宝石、蓝宝石、变石等，会严重影响其价格。

处理

宝石的处理如加热、裂隙充填、辐照和扩散会极大地提高很多宝石的外观。现在这些处理已经成了商业级宝石的例行工序。处理过的宝石总是比类似未处理的宝石便宜。但按惯例大多数宝石都会被处理，如红宝石和蓝宝石都经过处理，未经过处理的很少见，市场价格也超出大多数顾客的承受范围。如果你非要买未处理过的宝石，也还是有很多选择的。许多流行的宝石比如碧玺、尖晶石、紫晶和石榴石几乎都未经处理。

产地

严格来讲，不论产自哪个国家和哪个地区，一块好的天然宝石就是一块好宝石。但事实上某些地区如缅甸、克什米尔、斯里兰卡、巴西出产的宝石在市场上的价格会高一些。很难说这种价格是否合理，尤其是有很多好宝石来自非洲。

时尚

一些宝石，比如蓝色蓝宝石，总是很时尚。但有些宝石会在短期内流行，使得价格上涨。近来我们注意到，中长拉石、硬水铝石就属于这种情况，也有很多人对金红石水晶很感兴趣。一些非常好的宝石，比如天然尖晶石，其价格实际上低于预期价格，因为有限的供给使得这些宝石在市场中没有被大规模推广。

供给链

宝石贸易是买卖交易，从宝石开采到零售，每个人都处在供给链中且都设法获利。宝石从开采者手中到最终消费者手中要经过很多次交易，经过的代理和分销商越多，其价格就越高。所以从不同的地方购买相同的宝石，价格可能相差 200% 之多。

主要参考文献

陈钟惠. 珠宝首饰英汉-汉英词典(上册,第三版)[M]. 武汉:中国地质大学出版社,2007.
陈钟惠. 珠宝首饰英汉-汉英词典(下册,第三版)[M]. 武汉:中国地质大学出版社,2007.
郭颖,余晓艳. 珠宝专业英语[M]. 北京:地震出版社,2006.
胡楚雁. 宝石次生包裹体的成因分类及表现特征探讨[J]. 桂林工学院学报,2004(1):28~31.
施光海. 论宝石包裹体的分类[J]. 宝石和宝石学杂志,2008(2):58.
张义耀,余平. 珠宝英语口语[M]. 武汉:中国地质大学出版社,2007.
中华人民共和国国家质量监督检验检疫总局,中国国家标准化管理委员会. GB/T 16553—2017. 珠宝玉石 鉴定[S]. 北京:中国标准出版社,2017.
中华人民共和国国家质量监督检验检疫总局,中国国家标准化管理委员会. GB/T 16552—2017. 珠宝玉石 名称[S]. 北京:中国标准出版社,2017.
中华人民共和国国家质量监督检验检疫总局,中国国家标准化管理委员会. GB/T 16554—2017. 钻石分级[S]. 北京:中国标准出版社,2017.
Donoghue M O. Gems[M]. 6th ed. Oxford:Butterworth-Heinemann,2006.
Read P G. Gemmology[M]. 2nd ed. Oxford:Butterworth-Heinemann,1999.